Transnational Oil

Studies in International Political Economy will present new work, from a multinational group of authors, on major issues, theoretical and practical, in the international political economy.

Already Published

The Political Economy of New and Old Industrial Countries
The East European Economies in the 1970s
Defence, Technology and International Integration
Japan and Western Europe
Tax Havens and Offshore Finance
The North–South Dialogue
The International Gold Standard

Forthcoming Titles

Dependency Transformed
Defending Europe in the 1990's

Transnational Oil

Issues, policies and perspectives

Zuhayr Mikdashi

Université de Lausanne
(École des Hautes Études Commerciales)

Frances Pinter (Publishers), London

First published in Great Britain in 1986 by
Frances Pinter (Publishers) Limited
25 Floral Street, London WC2E 9DS

British Library Cataloguing in Publication Data
Mikdashi, Zuhayr
Transnational oil: issues, policies and perspectives.—(Studies in international political economy)
1. Petroleum industry and trade—Political aspects 2. International economic relations
I. Title II. Series
338.2'7282 HD9560.6
ISBN 0-86187-549-4

Typeset by Joshua Associates Limited, Oxford
Printed by SRP Ltd, Exeter

Contents

List of Tables

List of Figures

Preface

This study attempts to identify and evaluate factors which have been responsible for shaping major developments in the world oil sector. Particular attention will be devoted to analysing the objectives and strategies of enterprises and governments. Chapter 1 focuses on corporate behaviour of major operators. Chapter 2 surveys key forces in the energy market and interprets issues of oil price management. Chapter 3 scrutinises the principal objectives and policies of oil-exporting states, and the scope for or limitations of concerted action in the petroleum field. Chapter 4 examines major objectives and policies of industrial countries in the energy field. Chapter 5 analyses certain financial aspects of the world oil business, including the relevance of finance to oil exporters and oil importers. Chapter 6 reflects on the role of oil in the development process. The concluding chapter attempts to summarise some of the principal issues and policies analysed in this study.

This book builds upon earlier works the author has published on the subject of oil and international economic relations. It has relied on primary sources for facts and figures and has been enriched by frank exchanges with executives and public officials from various parts of the world. In writing this book the author has been guided solely by the pursuit of knowledge in a complex and often controversial field. He alone bears the responsibility for its contents.

The author acknowledges Dean Charles Iffland's support and the congenial academic atmosphere of the University of Lausanne's Business School (École des Hautes Études Commerciales, HEC), where the book was written.

The author appreciates discussions he has had with Susan Strange, Professor of International Relations at the London School of Economics, who is hosting the publication of this book in her series 'Political Economy Studies'. Equal appreciation is also due to Samuel L. Hayes III, Professor of Investment Banking at Harvard Graduate School of Business Administration, who has produced jointly with this author an essay on oil exporters' financial flows. A few researchers affiliated to industry, financial institutions or governmental organisations have given their feedback on earlier drafts of selected sections of this study. While it is impossible to acknowledge their valuable help (often to respect their wish for discretion), I hope they will recognise my sincere debt of gratitude. Finally, the author gratefully notes his wife's solicitude.

Lausanne, Switzerland *Zuhayr Mikdashi*
September 1985

1 Corporate Strategies

FACTORS OF CHANGE AND OPERATORS' OBJECTIVES

Natural crude oil or petroleum, in the narrow commercial sense, is defined as a liquid hydrocarbon (an organic compound of hydrogen and carbon) at normal temperature and pressure. Other hydrocarbons are solid or gaseous; for example, coal and natural gas, respectively. Most crude oils contain associated gas and solid hydrocarbons (e.g. asphalt) in solution. Petroleum was used in ancient times for heating, lighting, road-making and building, most notably in Mesopotamia, Egypt, Persia and China.

The modern petroleum industry's origin is often traced to Edwin L. Drake, who sank an exploratory well in 1859 near Titusville, Pennsylvania (USA), and struck oil on 30 August of that year at a depth of 695 feet. However, historical evidence shows that oil was already produced in Russia by 1856 and in Romania by 1857.[1]

Commercial operators, under US law, obtain mineral rights to exploit the subsoil riches of private land. When an oil-field extends over several private lots, each owner has the incentive to 'capture' as much of the oil riches as possible before they are taken by neighbouring operators. Thus, the discovery of new oil-fields—for example, in Pennsylvania (1859), Oklahoma (1903) and Texas (1920s)—provoked a spate of drilling and production. In the absence of laws imposing rational, efficient exploitation of oil-fields, highly parcelled-out land ownership thus led to wastefully dense drilling and excessive production of oil and gas. The rapid dissipation of reservoir pressure therefore eventually reduced the ultimate recovery of commercial reserves. Laws requiring field 'unitisation' to ensure optimal production by competing operators were first passed at state level in the United States in the 1930s.[2]

In the early years of the industry the behaviour of producers and refiners—often one-man operators in the United States—led to sharp swings in supplies and prices: oil-men resorted to price cuts as they discovered prolific reserves; and shortages occurred as exploitation and development slackened and wells dried up, or as demand surged. All these factors were responsible for acute price volatility.

A swing away from a market of numerous, unrelated and fiercely competitive operators started in the 1870s. Thus, thanks to a judicious strategy of cost-cutting rationalisation, railway and pipeline rebates, predatory pricing, acquisition and expansion, John Davidson Rockefeller sen.

(1839–1937) achieved virtually full control of refinery and later of oil transport facilities in the United States by the end of the nineteenth century. In 1882 he brought all his companies under the aegis of Standard Oil Trust, with headquarters in New York.

However, Rockefeller's business strategy aroused the bitterness of independent oil producers. The latter managed to rally sufficient popular and legislative support for various legal actions, notably through the passage of the Interstate Commerce Act (1887), the Sherman Anti-trust Act (1890), and various state laws imposing 'common carrier' status on pipelines and forbidding discrimination by railway or pipeline transporters.

In 1892 the Ohio Supreme Court ordered the dissolution of the Standard Oil Trust. Rockefeller consequently amended the charter of one of the Trust companies—namely, Standard Oil Company (New Jersey)—to make it a holding company for all his interests. The US Government charged the new organisation with restraint of trade in a suit brought in 1906, and also accused the holding company of having violated the Sherman Act. In 1909 the US District Court in St. Louis ruled against the Standard Oil Company (New Jersey) holding. The climax was reached on 15 May 1911, when the US Supreme Court, on the basis of the Sherman Act, issued a decree calling for the break-up of Standard Oil into thirty-four independently owned and managed companies.[3] Twelve of the successor companies had domestic marketing operations; some of these sold specialty products only. Nine were full-line marketers, initially with non-overlapping territories. Some of the successors later moved into marketing, refining, or producing in more than one state. Before its break-up in 1911, Standard Oil controlled 84 per cent of domestic refinery output, 85 per cent of domestic sales and 86 per cent of oil exports from the United States. By 1914 US oil production accounted for some 63 per cent of the world total.[4]

In the 1920s the former Standard Oil companies became genuine competitors and moved into each others' territories. Competition stirred up by large oil discoveries and exacerbated by the Depression of the 1920s led to 'rock bottom' prices and trade losses. To remedy their discomfiture, major US and European companies sought, by private agreement reached on 17 September 1928, to control sales and administer prices in international markets.[5] In this agreement, two major principles were espoused:

- exploration for, and development of, new oil-fields was to be made by way of partnerships among the contracting parties;
- the geographical participation of markets among contracting companies was to be maintained as it was then in subsequent years (the 'As-Is' agreement).

US anti-trust investigations, coupled with the revival of demand and the outbreak of hostilities in the Second World War, rendered the 'As-Is' agreement largely inoperative. Nevertheless, the world petroleum industry emerged after the war with a strong oligopolistic structure, largely dominated by some seven major companies. These companies benefited considerably from a vigorous expansion in demand for oil. In due course, 'newcomers' were enticed into the world oil industry, attracted by the relatively high profits and growth potential.

Significant changes are continuously occurring in the sources of oil production and their relative importance, as well as in the composition of operators. Nationalisations, divestments and mergers, as well as the exhaustion of oil-fields in certain countries and discoveries in others, have been responsible for changing patterns of supply. Government policies and measures have often played a major role in constraining or encouraging the development of, and trade in, energy products. Thus, when the United States became a net oil importer in 1948, it started to strengthen restrictions on crude-oil imports, beginning with a 'voluntary' programme and terminating with mandatory controls in 1959. Import restrictions were later relaxed, in the 1970s.

The world-wide expansion of the industry since 1860 has been spectacular with respect to both its size and influence. Equally remarkable has been the metamorphosis of government–company relations, especially outside the United States. Prior to the Second World War, several major oil concessions were obtained by sophisticated Western enterprises in countries whose governments had limited access to know-how; a few of these countries were under the tutelage of Western powers. The concessions provided the companies that acquired them with a number of wide-ranging advantages. In particular, these comprised:

- Large areas for exclusive operation, extending often over practically the entire territory of some countries, such as Kuwait, Iraq, Qatar, Bahrain, Oman, the United Arab Emirates, or over vast tracts in the larger countries such as Iran or Saudi Arabia. Often, provisions for the relinquishment of unexplored or undeveloped areas were not included or were left to the discretion of concessionaires.
- Extremely long duration, stretching over several decades, with terms including fiscal terms, fixed rigidly.
- Total discretion for the concessionaire with respect to the size, pattern and direction of its activities (covering exploration and drilling programmes, production levels, transport, refining, exports, etc.).
- Freedom in the import of goods and the recruitment of expatriates, in the construction and management of separate housing compounds for their employees equipped with schools, shopping and recreation centres, hospitals and other amenities.
- Settlement of disputes outside the jurisdiction of the competent national courts of host countries, through recourse to international arbitration.

As a *quid pro quo* for these advantages, the concessionaire agreed to make what turned out to be modest royalty payments. The royalty rate was fixed over the whole life of the concession, and little or no flexibility was granted to the host government to allow it to raise the rate in response to the discovery of low-cost prolific oil-fields and/or in the event of an increase in oil profits.[6] Royalty rates were based on either a unit of production or export, or the posted price of oil, itself fixed unilaterally by the concessionaire. In a highly concentrated and vertically integrated industry with no competitive market for oil, the concessionaire could set the levels of production and exports of oil and their prices with a measure of discretion. This discretion implied that the concessionaire had a say in varying the receipts of host countries. Several host countries were affected by that discretion in so far as they were heavily dependent on petroleum to finance the activities of the state and to develop the national economy.

One scholar of industrial organisation commented on the pre-1970 Middle East oil concessions thus: 'Never in modern times have governments granted so much to so few for so long'.[7] Comparing Middle East concessions with US practices, he noted:

> The prodigal nature of the Middle East concessions can be more readily appreciated by comparing them with the United States' policy in granting oil leases on public lands. Our federal policy has been designed to prevent monopoly exploitation of oil lands. Federal laws limit the area covered by exploration permits (not more than 246,080 acres in any one state, except Alaska, where the limit is 300,000 acres). They limit the area a prospector may lease after the discovery of oil to one-fourth of the lands embraced in the prospecting permit with a minimum of 160 acres. They limit the terms of a lease to twenty years. They require that three-fourths of the acreage covered by an exploration permit, to which the original lessee has a preferential right, be leased by competitive bidding with a minimum royalty of $12\frac{1}{2}$ per cent. They require that lands located in a known geological structure of a producing field be leased by competitive bidding in units of not more than 640 acres [1 square mile/2.59 square kilometres]. They limit the duration of a lease resulting from competitive bidding to five years or as long as oil or gas is produced in paying quantities.[8]

This is, however, a retrospective view. Prospects did not look as attractive when the very first concessions were granted, and companies often did not know whether oil reserves would be found in commercial quantities. A comparison with the United States after much oil had been produced is therefore not fully appropriate. Nevertheless, the first lease, granted at the end of 1857, before any oil was produced in the United States (see Figure) appears to give the American private lessor (by comparison with the Middle East government lessors in the first half of the twentieth century) attractive terms with reference to royalty or share in net profits—given the limited size of land leased, the short duration of the lease and the full uncertainty of

The Original Lease

Dated December 30, 1857
Deed Book P, p. 357
$1 in hand.

Pennsylvania Rock Oil Company
to
E. B. Bowditch and E. L. Drake

'Demise and let' all the lands owned or held under lease by said company in the County of Vanango, State of Pennsylvania, 'To bore, dig, mine, search for and obtain oil, salt water, coal and all materials existing in and upon said lands, and take, remove and sell such, etc., for their own exclusive use and benefit, for the term of 15 years, with the privilege of renewal for same term. Rental, one-eighth of all oil as collected from the springs in barrels furnished or paid for by lessees. Lessees may elect to purchase said one-eighth at 45 cents per gallon, but such election, when made, shall remain fixed. On all other minerals, 10 per cent of the net profits. Lessees agree to prosecute operations as early in the spring of 1858 as the season will permit, and if they fail to work the property for an unreasonable length of time, or fail to pay rent for more than 60 days, the lease to be null and void.'

Source: cited in Nordine Ait-Laoussine, *Exploration and Production Agreements in Developing Countries*, International Energy Development Corporation (IEDC), Geneva, December 1982.

discovering oil in the United States. Government levies, in the United States, it is important to note, come over and above the royalty or profit payments made by the lessee to the private lessor.

Prompted by the objective of protecting national interests, host governments of developing countries have since the Second World War sought to assume effective control over their domestic oil sector. No concession as originally contracted remained intact until its expiration date. Changes occurred in fiscal charges, in operational obligations, in control of production and export, in relinquishment of unexplored areas, in maturity periods of contracts, and in other terms. New principles and rules have evolved over the past years, and the terms of early concessions obviously look anachronistic today. Chief among the several factors which contributed to the improved bargaining stand of host countries in developing areas are:

- — political independence and growth of nationalism;
- — increased sophistication and civic consciousness of national decision-makers, coupled with their easier access to independent international expertise;
- — the desire to emulate governments in developed countries which have obtained attractive terms for oil leases on their public land (notably the United States);
- — the ability to invite competitive bidding from, for example, 'independent' companies;
- — the exchange and sharing of information and experiences, and concerted joint efforts through groupings of a regional or interregional nature—for example, the Organisation of Petroleum Exporting Countries (OPEC)—to improve terms;
- — the higher scarcity or market value of oil, notably after 1969.

Since the 1940s, the world oil market has been largely dominated by a few major transnational companies—Exxon, Mobil, Texaco, Chevron, Gulf, British Petroleum, and Royal Dutch/Shell. Chevron acquired Gulf in 1984 (see Chapter 5). Analysts often include other companies among the major transnationals, notably, Amoco, and Total (Compagnie Française des Pétroles).

Until the 1960s, major transnational companies managed, thanks to their large concessions and inter-company co-operation, to set effective barriers 'upstream' to the entry of competitors. Moreover, joint ventures among concessionaires or long-term contracts enabled the parties concerned to 'order out' their crude-oil liftings and to be aware of their production partners' intentions with respect to sales. Such business interrelationships among competitors are conducive to the exercise of a measure of restraint in business rivalry in product markets.

The erosion of the majors' market control has grown, with host governments opening up their uncommitted territories to competitive bidding by other companies. The bulk of these companies falls within three main groups:

— A few medium-sized companies, also known as mini-majors. These companies are predominantly US firms. They started in the 1950s, looking for cheaper foreign crude oil to replace declining or increasingly expensive national production to feed their refining marketing outlets in the United States, in order to compete on more equal terms with the majors which already had low-cost foreign crude supplies.
— European and Japanese state-owned or state-supported companies. Among others, the following countries have sought in this way to reduce dependence on the majors, for economic and national security reasons: France (Total and Elf-Aquitaine), Italy (ENI), West Germany (Veba Oel), Japan (Japan National Oil Corporation). (See also Chapter 4, section on 'governmental agencies'.)
— Companies from developing countries—oil exporting or oil importing. These companies have tried to develop national or even foreign oil resources, mostly since the oil price increases of the 1970s. Their activities were often launched with the nationalisation of domestic foreign-owned assets or through participation in joint ventures with firms from industrial countries.

Although companies vary considerably with respect to origin, size, structure, strategy, management, innovativeness, flexibility, performance, and other criteria, a few objectives offer a better basis for interpreting their behaviour. These are the objectives of high profits and growth, and the stability and long-term survival of the enterprise.

The main objective of any commercial enterprise is to produce adequate earnings for shareholders, taking into consideration the risks of the business and the opportunities available to investors. This primary concern should be compatible with the company's responsibilities towards its employees, customers, suppliers and society at large.

In market economies, managements of companies have naturally given primacy to shareholder interests. The chairman and chief executive of Gulf acknowledged, however, that 'shareholder interests are sometimes more honoured in word than in deed'. In so doing, management could be prompted by more selfish interests, most notably independence, tenure of office, high remuneration and power. Gulf's management resisted such temptations, its chief executive argued, when it fought the 'hostile, unfairly priced takeover bid by an investor group, led by Mesa Petroleum'. Gulf's board sought a 'fair' merger with Chevron, in the interests of the shareholders, and declined to 'save management—or to stay independent at any cost'.[9]

Principal strategies that have guided the behaviour of management in structuring and orientating their corporate activities are examined in what follows.

VERTICAL INTEGRATION

Oil company management has often sought the vertical integration of business functions; namely, a control of activities extending from the production of raw materials to their transport, transformation and distribution to end users. Management's drive for vertical integration is prompted by a number of objectives, chief among which are:

— the enhancement of the security of flows from sources of supply to outlets;
— the reduction of transaction costs;
— the achievement of informational advantage which permits rational programming of investments and activities.

The fulfilment of these aims should, in principle, reduce costs and enhance profit prospects. Thus, executives of a number of companies have considered it vital for their optimal long-term performance to establish a balance between the different activities of the oil business. Their aim is to reduce the risks and costs associated with imbalances between various company facilities, stretching from the sources of raw materials to market outlets, transport and refining. Indeed, shortages have to be filled, often at exorbitant rates, by recourse to outsiders, and surpluses have to be dumped at sacrificed prices in the open market.

Captive marketing outlets make marketing more profitable or less costly for companies which have crude-oil production assets, especially in periods of surplus productive capacity and keen competition. Symmetrically, equity crude available at favourable cost will make long-term procurement more attractive to companies that are refiners and distributors of petroleum products, especially in periods of shortage or restraint on crude-oil sales.

Vertical integration could start with 'crude-rich' companies moving into 'downstream' operations: namely, refining, distribution and chemicals. This was the case with British Petroleum. Vertical integration could move from 'downstream' operations (e.g. chemicals) into 'upstream' ones (i.e. exploration and the production of crude oil and coal). This was the case for the chemical company Du Pont, which bought an oil-, gas-, and coal-producing company, Conoco, in 1981 (the second-biggest coal producer in the United States and the country's ninth-biggest oil company) for $7.4 billion. Coal, gas and oil are important sources of feedstock for the chemical industry.[10] Vertical integration has also occurred in oil companies that have moved from being mostly in the import–export trade (general, including oil) into petroleum transport and other complementary operations. A notable example in this category is that of the 'Shell' Transport and Trading Company, led by Marcus Samuel, who in 1907 merged his company's assets with those of the Royal Dutch Petroleum Company, led by Henri Deterding (with capital shares split 40 per cent and 60 per cent, respectively). Royal Dutch had discovered oil in Sumatra in 1890, and in 1903 was operating a refinery in Rotterdam.

One needs, however, to be cautious in comparing companies' strategies across different historical contexts. One should add here that Shell's merger with Royal Dutch in 1906 followed on the heels of frequent price wars (from 1897 to 1905) and was prompted by the two companies' shared wish to withstand the fierce competition of the giant Standard Oil Trust.[11]

The relative decline in the major companies' vertical integration and market control in the 1970s and 1980s has been paralleled by a symmetrical growth of 'national' companies. The activities of the latter are largely based on producer or consumer countries. Moreover, they generally benefit from the active support of their public authorities and abide by their directives. By comparison, 'independent' companies are privately owned and do not generally have any special relationships with home or host governments. Their activities are largely concentrated in one function of the petroleum industry, such as crude-oil production, transport, refining or marketing. 'Independents' comprise a wide and diverse spectrum of operators, stretching in size from those with a multi-billion dollar turnover to the modest service station owner.

The deintegration of major oil companies after the nationalisation or take-over of their concessions in leading oil-exporting countries in the 1970s could not leave these companies indifferent. These companies have sought to reintegrate upstream either through new exploration ventures outside the OPEC countries or through the acquisition of other companies which control significant commercial oil reserves located in reasonably accessible regions of the world. Indeed, the 1984 take-over strategies of Chevron Corporation (previously Standard Oil of California—Socal) and Mobil were based on the assumption that investment in the acquired company would bring them cheaper crude-oil reserves than would be the case with investment in new oil exploration.[12] Thus, Chevron's acquisition of Gulf, at the cost of $13.3 billion, brought Chevron 2 billion barrels of net proved reserves at a cost of around $4.5 per barrel.[13] Mobil's acquisition of Superior, at the cost of $5.7 billion, brought Mobil net proved reserves of 1.07 billion barrels at a cost of less than $6 per barrel.[14]

The history of major companies shows an underlying and continuous strategy aimed at achieving an adequate degree of vertical integration. Notwithstanding legal and administrative constraints, these companies have managed to achieve or re-establish a large measure of vertical integration in their petroleum operations. One could illustrate the legal constraints encountered by major oil companies by reference to the US Supreme Court's order to break the Standard Oil Trust in 1911, and the wave of nationalisations of major oil companies' interests in oil-exporting countries (e.g. in Mexico before the Second World War and in several OPEC countries in the 1970s).

GEOGRAPHICAL SPREAD

Major petroleum enterprises have built extensive networks of affiliates, subsidiaries, joint ventures, and so on, over various parts of the world. This strategy of geographical spread enables the company concerned to exploit business opportunities rapidly; to make up for an unexpected disruption of supplies, transport or refining facilities in a given region of the world by stepping up activities in other regions of the world; and to reduce over-dependence on one region or one market, thus attaining greater corporate stability—with one market's profits helping to mop up another market's temporary losses. Besides these motives, geographical spread contributes to building, in the minds of users, perceptions of ready availability of its products and services. In consequence, this feature enhances consumers' loyalty to the enterprise concerned.

Geographical extension can be achieved through the creation of new operational units in competition with existing enterprises. Alternatively, geographical extension can be realised through acquisition of, or merger with, existing enterprises. The latter strategy has often been favoured by major enterprises keen to establish an immediate position in a new market, in the absence of legal constraints. By comparison, the building-up of new affiliates from 'scratch' could be too time-consuming.

Concerned about spreading risks and investments in different countries and projects, and/or exploiting business opportunities which are too costly to be undertaken by any one single enterprise, oil companies have resorted to joint ventures in one or more of the functions of the industry. These could cover exploration and production, transport, refining, manufacturing or marketing in various parts of the world. Because of their interrelatedness and co-operation in such joint ventures, the major transnational energy enterprises have been dubbed the 'sisters'. Although they have seldom managed to float sustainable cartel organisations, their policies *vis-à-vis* governments and consumers have often been characterised by convergence—largely because of mutual interest in the market-place.

A company can obtain fiscal advantages, thanks to its transnational spread, by crediting losses of one affiliate against the profits of another and by setting 'appropriate' transfer prices for raw materials, processed products, services, technology, and so on. Indeed, in the absence of 'arm's length' open-market prices and barring legal interdictions, the transnational vertically integrated company is naturally prompted by a wish to price inter-affiliate transactions in such a way as to reduce its overall tax burden, by attributing the lowest possible profit margins in high-tax countries and the highest possible profit margins in the low-tax or tax-free countries. Such manœuvrability is not accessible to the 'independent' companies: their activities are mostly limited to one country and/or one kind of operation (for example, production, transport refining or marketing). As one US source pointed out:

> By charging unrealistic prices in the exchange of goods and services, a corporation can transfer profits earned in one country to another. For example, an artifically high transfer price for goods or services purchased from a foreign subsidiary can reduce a parent US corporation's profits and, thus, the tax revenue to the US.[15]

The above assumes the absence of adequate public information on competitive market prices. This was the case in the world oil industry until the 1960s. Since then, a significant 'open' market evolved, spearheaded by Japanese and Italian independent refiners. The growth of that market, along with the dissemination of public information on prices and the vigilance of fiscal authorities, has eliminated the scope of discretionary pricing by transnational corporations.

SIZE

The structure of the international energy market has been characterised in this century by the presence of a number of large and diversified enterprises. Energy companies in the modern industrial era are typically high risk and capital intensive. Accordingly, size is a factor of considerable significance in achieving:

- distribution of risks (notably exploration risks) among a large number of ventures to have a reasonable chance of success;
- economies of scale and rationalisation of operations in a typically capital-intensive industry; these operations comprise, *inter alia*, drilling, pipeline or tanker transport, refinery and petrochemical plants;
- diversity of products and services, thereby providing customers with the convenience of meeting multifarious needs under 'one single roof'.

Indeed, oil is a risky business at the exploration, political and commercial levels. Considerable expense is sustained in exploration programmes, some of which could prove to be unproductive. This necessitates the spreading of exploration over several distinct tracts of land in different countries. To illustrate, BP and its affiliate Sohio spent several hundred million US dollars in 1982–83 but by 1984 had not discovered any giant fields (comparable to their 1970s finds in the North Sea and Alaska) which would replace their declining reserves. A particular well off North Alaska turned up dry by the end of 1983 after costing BP and other companies as much as $1.6 billion in drilling and other related exploration costs.[16]

Besides geological risks, commercial losses can be incurred following inopportune fluctuations in prices. Moreover, political factors—such as boycotts/embargoes, nationalisations or conflicts—can be causes of large losses. Large size thus enables a company to absorb such losses without jeopardising its survival. Large size can also mean that adequate resources (human and material) are available for new opportunities which can fructify over the longer term.

If judiciously used, size can serve the interests of the company (under certain propitious circumstances) by improving its bargaining position *vis-à-vis* its competitors, customers, suppliers and the public authorities. Lobbying has been one of the approaches used by business to influence to their advantage law-makers and law-enforcers. In the United States, both big business and small business have vied for the favourable attention of the Administration and the legislature. As one scholar put it:

> The strength of the independents, then [in 1911 when the US Government enforced the break-up of the Standard Oil Trust] as now [1975, date of article], rested in part on the fact that they were well distributed over the face of the United States and could rally formidable Congressional support for any position they took.[17]

Various approaches have been used in the oil industry to achieve size: they include vertical integration, take-over, joint ventures, diversification into other sources of energy and, to a much lesser extent, diversification into other business sectors. One approach favoured by Japanese industry for obtaining the advantages of size has been through 'tie-up' arrangements, aided by the prodding of the Ministry of International Trade and Industry (MITI). This was the case with Nippon Oil and Mitsubishi, which agreed, initially for the period 1985–89, to link up crude-oil purchases, transport, storage, oil-product refining and distribution. The combined share of the two companies in Japan's fuel-oil products was, towards the end of 1984, 25.1 per cent. The two companies expect to save some $40.4 million a year by rationalisation measures: this does not take into consideration other advantages derived from their improved competitive (i.e. notably bargaining) position in obtaining crude-oil supplies on more favourable terms.[18]

One oil group, the French state-owned Elf-Aquitaine, praised the advantages of size and supported its realisation in the United States via such a strategy as mergers and concentration of petroleum companies.[19] How large is 'big enough' remains an open question. Indeed, the ten largest firms comprise quite a range of sizes. Nevertheless, the adequacy of size should be related to the functions undertaken with a view to achieving economies of scale and efficient rationalisation of operations, within the bounds of a competitive market structure.

SECTORAL DIVERSIFICATION => horizontal integration ?

Diversification outside the oil sector has often been prompted by a set of complex factors, chief among which are:

— perceived attractive business opportunities, currently or prospectively;
— moderation in the fluctuations (cyclical or otherwise) of business activity resulting from excessive dependence on one line or one sector;
— fuller development of complementarities at various business levels, such as those of production, transport, marketing and know-how.

The management and owners of an enterprise do not normally wish to fold up business with the drying-up of a supply or the obsolescence of a product or service. This principle is also valid for a petroleum company. As commercial sources of energy are variable over time, rational management calls for changing the mix of products with due regard to business opportunities. Thus, to protect themselves against a narrow dependence on one commodity, several oil companies have moved since the early 1970s into so-called 'non-traditional business areas'. According to a chief executive of Europe's largest oil company, this is 'to complement existing expertise and enhance long-term prospects for profitability'.[20] These areas cover energy outside hydrocarbons, chemicals and minerals, as well as other businesses.

This policy of diversification can best be portrayed by quoting the chief executives of the world's biggest oil company. They said:

> A decade ago Exxon's investments were almost entirely concentrated in traditional oil and gas and related petrochemical businesses. Today it is well on the way to becoming a broadly based energy company. It has substantial reserves of coal, oil shale and uranium, and important investments in oil-sands operations and synthetic fuels programs. It has a stake in solar energy and is pursuing research on other renewable energy forms. The 1979 acquisition of Reliance Electric Company furthers its commitment to the development of energy-saving technologies.
>
> The corporation has also been gradually extending its earnings base beyond the energy field into industries where its managerial and technical resources are particularly relevant. Thus, Exxon is pursuing opportunities in copper and other metals and in selected high-technology businesses with significant future potential, such as those related to information systems.[21]

Several oil companies have followed the path of non-energy diversification, either in directly developing new activities or through the acquisition of already established enterprises. In the area of acquisitions of minerals/metals enterprises, notable examples are those of Shell with Billiton (1970); Atlantic Richfield Co. (ARCO) with Anaconda (1977); British Petroleum with Selection Trust (1980); Standard Oil, Ohio (for short, Sohio—in which BP holds a majority interest) with Kennecot, the world's biggest privately owned copper producer (1981); Elf-Aquitaine with Texasgulf (1981).

Other oil companies were not able to acquire mining/metal enterprises either because the value of public offers made for the shares of the vied-for company/companies were considered inadequate by the management or shareholders of those companies, or because of objections by the regulatory authorities. Thus, Chevron Corporation (originally Socal) failed on two occasions in 1981 to acquire Amax. As an alternative to the take-over approach, several oil companies have resorted to participating in the equity of independent mining enterprises and/or directly to launching large exploration programmes.[22] One such company was Amoco, whose affiliate in mining had $1.8 billion in assets in 1984.

Beyond energy and minerals/metals, there are a few important cases of oil companies' acquisition distantly related or unrelated to their core business. This is notably the case for major as well as for independent companies. A major British oil company, BP, has thus not only diversified into other sources of energy chemicals and into the minerals/metallurgical industry, but it has also moved into nutrition, detergents and information technology.[23] In 1981, a US oil company, Occidental, acquired the biggest US beef-packer for $800 million.

Between 1976 and 1980 Exxon fell for the strategy of diversification in business sectors unrelated to its basic energy activities by buying high-technology companies (Vydec and Reliance Electric Co.). These acquisitions proved too costly. As a result, Exxon decided in 1985 to divest some of these assets (notably Exxon Office Systems) and to concentrate on its energy business.[24]

As a further illustration of diversification outside its core area of business, Mobil (the third-largest US oil company) acquired, in 1979, important companies in the trading and transport sectors for $1.8 billion, namely, the Montgomery Ward shopping centres chain (sixth-biggest US retailer) and Container Corporation of America (the biggest US paper-packaging group). The motivation, as explained by the president of Mobil (W. P. Tavoulareas), was: 'We diversified in the United States because we were worried somewhat about our future in the United States. . . . We were worried, with all these bills in Congress, all of these regulations, as to whether we could make a fair return in the United States'.[25]

In 1981 Mobil even contemplated (but did not realise) a bid for some 25 per cent of the stocks of the US Steel Corp: this move was deliberated shortly after US Steel obtained a majority holding in an 'independent' oil producer (Marathon Oil) at a cost of $6.4 billion, after defeating Mobil's bid.[26] By 1985, owing to the poor results obtained from Montgomery Ward, Mobil's chairman hoped to divest that affiliate.[27] A US major, Texaco, acquired non-oil assets by taking over Getty Oil Company in 1983 for $10.1 billion. The non-oil assets were insurance, entertainment and sports, and agricultural-products properties. Texaco's objective, however, in acquiring Getty was that of 'increasing crude-oil reserves'. It later sold some of Getty's non-oil assets.[28]

It is doubtful whether size, achieved through the acquisition of non-related businesses, is an efficient allocation of capital resources and, accordingly, whether it is in the longer-term interest of stockholders and the national economy. Indeed stockholders could be faced with an involuntary stake in a different line of business whose acquisition cost could be 'excessive' and whose profitability could prove disappointing.

While one should acknowledge that not all diversification moves by the oil industry have been failures, several factors have been responsible for the poor performance of oil companies in their sectoral diversification. Chief among these factors are:

— unfamiliarity with key commercial or technical aspects of the new business;
— underestimation of the extent to which the staff of the acquired company have been demotivated;
— poor management;
— high cost of the acquisition;
— the need to revamp and sustain the acquired business by additional substantial infusion of capital (e.g. for renovation, modernisation or environmental reasons);
— deep and protracted recession affecting the world economy.[29]

KNOW-HOW AND LOGISTICS

Unlike a corporation whose activities are limited to one country, a transnational corporation, by definition, operates in several countries. It thus represents a cluster of enterprises in diverse legal jurisdictions, all sharing common ownership, objectives and strategies. Each transnational corporation engages with its foreign affiliates or subsidiaries in a variety of sale–purchase transactions, covering raw materials, finished or semi-finished products, machinery parts, licences, technology and other products or services. Such transactions call for efficient lines/methods of communication, co-ordination and co-operation.

Innovation in the area of logistics and information has been remarkable, aided by the computer–telecommunication revolution. Thus, it has become feasible, though not yet operational, to integrate a local distribution point (e.g. the underground petrol tanks at a service station) to a refinery and an oil-field. This system boosts operating efficiency in so far as it establishes a chain of links between the stocks available for the ultimate customer and the requirements in transport, refining and production needed to satisfy variations in the level and composition of demand.[30] Another illustration is the linkage commonly used between offshore production platforms and onshore pipeline distribution centres by means of satellite telemetry.

The energy market has been changing fast. To keep up in the leading group of enterprises, know-how is crucial. This requires:

— efficient management of complex organisations whose size and transnational spread call for particular skills, including those of reconciling, balancing and managing competing or contradictory interests;
— a matching of resources with rapidly changing requirements, especially when the lines of communication and flows are long and subject to unforeseen shifts or interruptions;
— continuous development of additional or alternative sources of products and services better adapted to changing needs, and the improvement of the efficiency of processes.

In this context one could illustrate the logistic capabilities of oil enterprises by referring to their efficient response in periods of crisis. This was the case during the Arab oil embargo applying to certain destinations following on the Arab–Israeli conflict of October 1973. It took oil companies virtually one week to rearrange the pattern of global sourcing, transport, refining and delivery of oil—to minimise the impact of disruptions.

One major constraint encountered by energy enterprises in the development of their activities is the limited availability of qualified personnel. Indeed, financial capability is not an insurmountable problem where managerial–technical talents are available. Energy enterprises, owing to the complexity of their activities, call for multi-disciplinary expertise. To recruit, train and keep first-class expertise constitutes for these enterprises an unremitting challenge.[31]

NEWCOMERS FROM OPEC

The companies which come closest (1985) to fitting the previously outlined characteristics are six transnational companies. These have dominated the world energy market since the Second World War. Their shareholding and top administration is largely American (Exxon, Texaco, Mobil and Chevron), Anglo-Dutch (Royal Dutch/Shell) and British (BP).

At the time of OPEC's creation in 1960, member countries' share in the production of their petroleum was extremely modest (about 2 per cent). In twenty years it rose to over 75 per cent. This dramatic increase resulted mostly from the purchase or nationalisation of oil assets (oil-fields, refineries, distribution networks), hitherto owned and operated by the major transnationals in OPEC host countries.

Among OPEC national oil companies, the one which has moved vigorously beyond national boundaries is the state-owned Kuwait Petroleum Corporation (KPC). The two primary objectives of the Kuwait government in this expansion were, first, to diversify Kuwait's oil activities world-wide, both upstream (exploration and production) and downstream (transport, refining, chemicals and distribution); second, to maintain or raise Kuwait's earning capacity, via judicious foreign investments in the absence of attractive national opportunities and to make up for the depletion of their national asset, oil. These objectives were commented on by Kuwait's oil minister and KPC's chairman thus:

> On the international level, also, we have been very active. Here we have two objectives. Firstly, to make investments that will augment the oil industry in Kuwait from the strategic point of view, and secondly, to make investments that will yield a higher rate of return. We have, for example, been successful in acquiring a substantial share (20 per cent) in Metallgesellschaft (over 1980–81), which owns Lurgi, and we hope to use this connection to improve the engineering know-how in the oil industry in Kuwait. We have been looking at other industries and companies that would bring advantages for our petrochemical industry, and in general we have been looking at projects which would combine a high level of profitability with opportunities to improve the overall strategic position of KPC.[32]

Early in 1982, KPC consummated a $2.5 billion take-over of a US major oil-exploration group, Sante Fe International Corporation, and by 1985 had completed the full take-over of another international oil-exploration group, the International Energy Development Corporation, based in Geneva.[33] By 1984, KPC completed the acquisition of much of Gulf Oil's assets in Western Europe, involving several refineries especially designed to treat Kuwaiti crude, and a network of about 3,000 service stations. KPC's exploration and production activities extend over the six continents of the world, in twenty countries.[34]

The Kuwaiti government oil-industry strategy is not limited to KPC. The Kuwait Investment Office (KIO) of the Ministry of Finance is another means of participation or control of petroleum or petrochemical ventures. Thus, KIO holds nearly 25 per cent of the equity of Hoechst, the West German chemical giant. This has facilitated the conclusion of an agreement between KPC and Hoechst for the sale of large quantities of Kuwaiti-produced petrochemical feedstocks.[35] KIO also has a large equity interest in British Hay's Group, with some 600 filling stations in the UK market.[36]

The growth and behaviour of KPC is reminiscent of that of Western energy companies under government ownership, such as Société Nationale Elf-Aquitaine (France), and Ente Nationale Idrocarburi (Italy). These companies have conducted their world-wide activities much like privately-owned enterprises and have sought functional and transnational extension. This was the case with Elf-Aquitaine, which, in July 1981, acquired a major US minerals company (Texasgulf) for $2.3 billion. The latter has important mineral reserves of phosphate, trona (sodium carbonate) and sulphur; it also produces fertilisers and nutrients derived from phosphates.[37]

One could consider Norway's state-owned oil company, Statoil, as being in the same league as KPC. Statoil activities span three continents: Europe, Asia and Africa. In the mid-1980s, furthermore, it is pushing for a complete chain of vertically integrated activities, from exploration for oil to sale of products.[38]

Notwithstanding these similarities, KPC has certain particular characteristics of its own:

— a shortage of Kuwaiti executives. This is understandable since Kuwait is a small developing country of 1.5 million inhabitants (about two-thirds of whom are expatriates);
— less bureaucracy at the top level, compared with those entrusted with routine work (this makes for rapid strategic decisions in such areas as major contracts, acquisitions or investments, whereas one can encounter undue delays in processing certain authorised transactions);
— as an entirely state-owned company in a capital-rich country, KPC is not faced with stockholder pressure to pay regular dividends. It can thus take a longer-term view on investments.

Besides the foregoing strategies, enterprises (oil and non-oil) have used 'expedient' approaches. By way of illustration, one may mention that sources in the Senate Select Committee on Intelligence attribute a key role to 'the Central Intelligence Agency [which] helped several American companies to undermine a contract that would have given the late Aristotle Onassis complete control over oil shipments from Saudi Arabia'.[39] Also, in 1976, a few oil companies openly acknowledged payments made to political parties in Italy over several years. Payments by BP and Shell totalled some £3.3 million. Among the Italian parties, the main beneficiary was the Christian Democratic Party.[40] These payments were not then deemed to be in violation of the law.

However, such corporate strategies represent aberrations and are not typical of the general behaviour of an industry concerned with prospering over the long term in harmony with the socio-regulatory environment of different countries. Evidence of these aberrations can be documented in countries where the transparency of information is greatest, notably in the United States and Western Europe.

NOTES

1. Royal Dutch/Shell, *The Petroleum Handbook*, 6th edn, Amsterdam, Elsevier, 1983, pp. 3–4.
2. Stephen L. McDonald, *Petroleum Conservation in the United States, An Economic Analysis*, Baltimore, Maryland, The Johns Hopkins Press, 1971, pp. 40–1.
3. See, for example, Amoco, 'Why Amoco Corporation?', *Torch*, Chicago, May–June 1985, pp. 2–3; and Albert L. Danielsen, 'The Standard Oil Refining Monopoly', in his *The Evolution of OPEC*, New York, Harcourt Brace Jovanovich, 1982, pp. 75–80.
4. See, for example, R. W. Ferrier, *The History of the British Petroleum Company*, 1, Cambridge, Cambridge University Press, 1982, p. 1.

5. US Federal Trade Commission Staff Report, *The International Petroleum Cartel*, Washington DC, 1952.
6. Venezuela is an exception in so far as it retained in the original concession agreement its freedom of action in matters of royalty and tax rates.
7. George W. Stocking, a one-time president of the American Economic Association, in *Middle East Oil, A Study in Political and Economic Controversy*, Nashville, Tenn., Vanderbilt University Press, 1970, p. 130.
8. Ibid.
9. James E. Lee, chairman of the board and chief executive officer, Gulf Corporation, 'Why We Sold to Chevron', *Leaders* magazine, New York, 1985, I, pp. 58–9; see also Chapter 5 below (pp. 000–00).
10. See, for example, 'Du Pont makes a Conquest, and America's biggest merger', *The Economist*, 11 July 1981, pp. 63–4.
11. See, for example, Ferrier, op. cit., pp. 1–5.
12. See testimony of Michael Pertschuk (Federal Trade Commissioner) in 'US Senate, Committee on the Judiciary', *Oil Merger Activity, Hearing*, Washington DC, 1984, 15 March 1984, pp. 49–50 and 61.
13. See Table 25 below (pp. 134).
14. 'Superior Oil Shareholders will vote September 20 on Mobil–Superior Merger', *Mobil World*, New York, August 1984, p. 8.
15. US Comptroller General, *IRS Could Better Protect US Tax Interests in Determining the Income of Multinational Corporations*, report to the Chairman, House Committee on Ways and Means, Washington DC, 30 September 1981, p. 2.
16. 'Sohio abandonne le forage pétrolier le plus coûteux du monde', *Le Monde*, 25 January 1984, p. 38; 'British Petroleum: Self-Reservation', *The Economist*, London, 21 April 1984, pp. 54–5; and Ken Wells, 'Oil Firms Pay $877 Million for Beaufort Leases', *Wall Street Journal*, 24 August 1984, p. 5.
17. Raymond Vernon, 'The Influence of the US Government upon Multinational Enterprises: The Case of Oil', in *The New Petroleum Order*, ed. Antoine Ayoub, Quebec, Canada, Université Laval, 1976, p. 56.
18. See 'Japanese oil majors in link-up', *Financial Times*, London, 4 December 1984, p. 16.
19. 'La nécessaire dimension', *Bulletin mensuel d'informations*, Elf-Aquitaine, Paris, March 1984, pp. 1–2.
20. 'Shell' Transport and Trading Ltd., *Annual Report 1980*, London, p. 12.
21. C. C. Garvin Jr (Chairman), H. C. Kauffman (President), Exxon Corp., *1979 Annual Report*, p. 3.
22. See also United Nations Conference on Trade and Development (UNCTAD), *The Promotion of Multinational Enterprises of Developing Countries*, Geneva, 16 May 1983 (TD/B/C.7/56), p. 26; and Banque Bruxelles Lambert, 'Metaux non ferreux: vers des horizons plus favorables?', *Bulletin Financier*, Brussels, 12 February 1982, pp. 1–6.
23. The Chairman's Statement, British Petroleum plc, *Annual Report and Accounts 1984*, London, p. 5; and *The Minerals Industry and BP*, Briefing Paper BP, London, February 1984.
24. 'Exxon 1985 Annual Meeting of Shareholders', *Exxon News*, giving 'Remarks to Shareholders by C. C. Garvin Jr, Chairman'; also Exxon *1984 Annual Report*, and 'Exxon: A Drive to Disaster in the Office', *Business Week*, New York, 3 June 1985, pp. 70–1.
25. *Domestic Crude Oil Decontrol, 1979*, Hearing before the Subcommittee on Energy and Power of the Committee on Interstate and Foreign Commerce, House of Representatives, 96th Congress, 1st Session, Washington DC, April–May 1979, p. 279.
26. 'Mobil Threat to Buy US Steel Stake is Bold or Frantic, Depending on Who is Speaking', *Wall Street Journal*, New York, 16 December 1981, p. 29.
27. 'Mobil tries to make the best of a bad buy', *Business Week*, New York, 20 May 1985, p. 23.
28. Chairman John K. McKinley Report to Stockholders, dated 12 March 1985, in Texaco Inc., *1984 Annual Report*, pp. 2–3.

29. Also M. R. Rendle, Managing Director, British Petroleum, *Managing Change, BP and the New Business Environment*, London, 8 June 1984.
30. See John Griffiths, 'Oil link to the local garage', *Financial Times*, 25 March 1982, p. 9.
31. Also P. B. Baxendell, chairman of 'Shell' Transport and Trading Ltd., in *The Human Resource*, Shell Briefing Service, London, No. 2, 1981, p. 2.
32. Ali Khalifa Al-Sabah, 'Oil Minister Outlines Strategy for Kuwait's KPC', in *Middle East Economic Survey* (*MEES*), *Supplement*, Nicosia, 18 May 1981, p. 2.
33. 'Kufpec Takes Full Control of IEDC', *Arab Oil and Gas*, Paris, 16 April 1985, p. 14.
34. Kuwait's Oil Minister Ali Khalifah Al-Sabah, in *MEES*, 28 May 1984, p. A6; and *KPC's Fourth Annual Report 1983–1984*.
35. 'Kuwait—Hoechst to Buy Petrochemical Feedstocks from Petrochemical Industries Company', *Arab Oil and Gas*, Paris, 16 May 1983, pp. 12–13.
36. John Cranfield, 'Kuwait—Cautious Downstream Strategy', *Petroleum Economist*, London, October 1984, pp. 380–1.
37. See 'Avec Texasgulf', *Bulletin mensuel d'informations*, Elf-Aquitaine, Paris, August 1981, pp. 1–2.
38. See Interview with Statoil President Arve Johnson, 'Statoil on three continents', in *Statoil 85/2*, Stavanger, Norway, pp. 4–5.
39. The US companies were Exxon, Chevron, Texaco and Mobil: 'Onassis Oil Deal Reportedly Upset with CIA's Help', *New York Times*, 19 December 1985, p. C8.
40. Ray Dafter, '£3.3m given to Italian parties in five years—BP and Shell admit political payments', *Financial Times*, London, 14 April 1976, pp. 1 ff; and 'Italian oil scandal—big deal', *The Economist*, 17 April 1976, p. 86.

2 Market Forces and Price Management

Growth in modern societies has been generally associated over the long term with the availability of useful energy at reasonable terms. This chapter seeks to identify factors which influence the conditions of supply and demand for oil, and consequently influence prices. An attempt is made hereafter to ascertain the role of market forces, companies and governments in the formation of oil prices.

VARIATIONS IN DEMAND—SUPPLY BALANCE

Forces influencing demand and supply are many and varied in nature. They can be grouped according to a number of categories; for example, economic versus political impact, or short-term versus longer-term impact. Often, participants in the world oil market are concerned about a measure of stability in oil flows and prices. If changes are likely, participants would wish to ascertain whether these are temporary or lasting.

In the short-term category one can include reversible factors whose influence is transient. These cover variations in weather conditions; breakdowns in flows from man-made or natural causes; 'overflows' owing to 'destocking' actions prompted by, for example, onerous interest charges or high expectations of impending price declines; 'distress' buying or selling; speculative transactions; and so on. Speculation, it is worth noting, can be ongoing in commodities markets. Shortages occurring in the short term, whether deliberately (man-made) or resulting from, say, accidents or disasters, can lead to panic-buying and can push prices to a level higher than the longer-term economic value of oil to the consumer or to a level higher than the cost of alternative fuels. Such price surges are possible owing to the high costs of switching over and the length of time needed to develop alternatives.

Factors of a longer-term nature affecting energy markets include those which lead to structural transformations whose influence is bound to be felt consistently over a number of years. These factors are generally prompted by expectations of lasting cost–price changes. Structural transformations affecting energy demand may cover a number of areas, notably:

— life-styles; for example, greater resort to public transport and/or to compact individual cars, and greater reliance on telecommunications which could save substantially on the need for travel;

— innovation in the field of technology, notably by application of advanced know-how to produce and/or consume various sources of energy efficiently;
— government policies, notably in the areas of fiscal burdens or incentives, the control or decontrol of domestic markets (covering policies in the fields of trade, prices, concentration, etc.);
— change in the mix of industrial output;
— levels of economic growth.

Price quotations for daily spot transactions on oil are readily sensitive to the shorter-term factors. Spot prices are more volatile than prices agreed on for contractual deliveries over specified periods.

Energy analysts are interested in measuring the effects of change in key economic variables (particularly price) of one fuel against the purchase or supply of other fuels. Using the concept of cross-elasticity of substitution, a measure can be made of the percentage change in either the demand or the supply of substitute energy products which results from an increase in the price of a given energy product, say petroleum. The change is reckoned to be limited in the short term since it takes time to replace equipment as well as to alter production and consumption habits. Some industrial plants are equipped with dual systems of energy generation (e.g. oil versus gas or coal). For these, the switch can happen very quickly.

Demand

The principal marketable product extracted from crude oil in the second half of the nineteenth century was kerosene, used for heating or lighting purposes. The take-off of the transport industry at the beginning of the twentieth century proved momentous for the petroleum industry. By 1914–15, gasoline had become the main petroleum product in demand. From the 1920s until the early 1970s, the demand for fuel oil grew rapidly and encroached on coal, formerly used in rail and sea transport as well as in factories. Oil consumption's growth in the twentieth century until 1973 was largely aided by its relative cost-attractiveness and its convenience with respect to transport and storage.

The demand for energy is a function of a number of factors, chief among which are per capita gross domestic product (GDP), population, energy prices, economic structure and the relative importance of energy-intensive activities, technological change with reference to raising the efficiency of energy-using equipment, and climatic conditions.[1]

The figures in Table 1 can help us understand the likely links between the price of oil and energy use. Assuming that consumers' prices for petroleum products have increased roughly parallel to those of crude oil,[2] oil price stability over the years between 1960 and 1973 appear to have been responsible for a 7.5 per cent per annum growth rate in world oil consumption out-

Table 1 Energy indicators for non-communist countries, 1960–1982 (compound average annual percentage change)

	1960–73	1973–82
Energy consumption	5.3	0.6
Oil consumption	7.5	−0.7
Real price of crude oil	−0.1	23.0

Source: The Chase Manhattan Bank, *The Petroleum Situation*, New York, August 1983, p. 1.

side the centrally planned economies. A price increase of 23 per cent per annum between 1973 and 1982 brought, by comparison, a decline in oil consumption of about 0.7 per cent per annum.

Higher per capita levels of energy consumption do not necessarily imply higher levels of waste. Moreover, a number of developing countries with low levels of per capita energy consumption have made inefficient use of their energy supplies. Energy use per unit of output is in principle a better indicator of energy efficiency than energy consumption per capita—assuming all other factors to be constant. In this respect, the analyst has to be careful to make appropriate provisos when comparing energy use in economies with different production–consumption structures; similar care is needed in comparing energy use over time in an economy which is passing through cycles of recession and rapid growth or through a process of transformation in its economic activities. Thus, an economy geared to 'know-how' industries is bound to use, per unit of output, less than an economy with an important energy-intensive industry (e.g. metallurgy). Nevertheless, it is worth noting that a developing economy such as China used two to four times more energy than major industrialised countries per unit of output in 1981. This is attributed to antiquated industrial boilers and other high-energy-using machinery.[3]

A lasting price increase in one source of energy should lead *ceteris paribus* to a shift from that energy source to its substitutes. Thus, increases in oil prices in the 1970s and perceptions of greater future uncertainty in oil supplies prompted users in seven major Western industrial nations to diversify the mix of their energy consumption in favour of non-oil sources (Figure 1)). Oil's market-share contracted significantly between 1979 and 1984 (Table 2). The sectors most affected were manufacturing, home heating and electricity generation. Oil use in electricity generation, for example, fell from 25.6 per cent in 1973 to 13.3 per cent in 1983 in the IEA countries.[4] Coal remains the most significant energy agent used after oil. Its share of world energy consumption is likely to rise among other fossil fuels since it has the bigger reserves. In 1983 coal still had 240 years of production, compared with thirty-five years for oil (see Table 5).

Table 2 Composition of world energy consumption (%)

	Oil	Natural gas	Coal	Hydro	Nuclear	Total
1979	45	18.4	28.5	5.9	2.2	100
1981	42.4	19.3	29.2	6.2	2.9	100
1983	40.3	19.2	30.3	6.8	3.4	100
1984[e]	39.3	19.7	30.3	6.8	3.9	100

[e] Estimated

Source: Union des Chambres Syndicales de l'Industrie du Pétrole, *L'industrie Française du pétrole 1984*, Paris, p. 28.

Assuming a general increase in all energy costs, energy demand per unit of GNP is likely to decline over the longer term, thanks to conservation and improved energy efficiency. Total energy consumption per unit of GNP fell by a factor of some 25 per cent between 1973 and 1984 in the seven major Western industrial countries. Improved energy efficiency covered all economic activities (extending from manufacturing to aircraft, vehicles, household durables, and so on) as well as to new technological applications (e.g. new electronic controls for heating and cooling systems). Industrialisation, growth in income, growth in population and other variables would have a stimulating effect on energy demand.

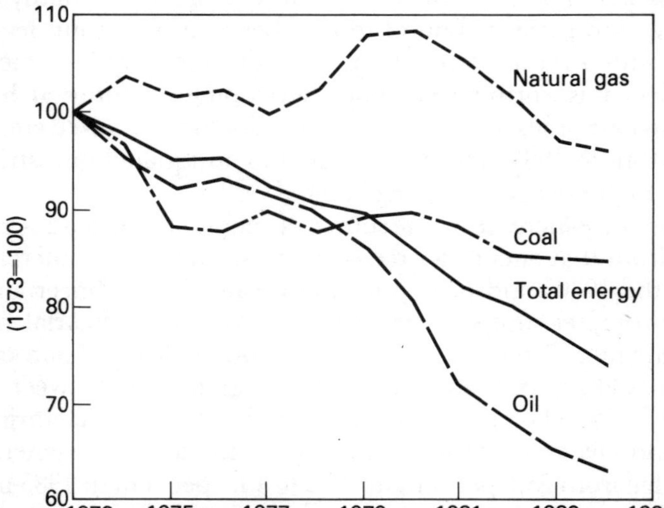

Figure 1. Energy per unit of GNP of the 'big seven' industrial nations.
Source: Texaco, *Free World Energy Review: Emerging Issues and Long-Term Forecast*, New York, 6 May 1985, p. 2.

Energy wasted by users may result from a number of factors, such as:

— lack of adequate insulation of buildings;
— inappropriate design of energy-using equipment, such as metal furnaces or transport vehicles;
— use of energy-intensive processes or products (e.g. metal containers instead of wood);
— sub-optimal organisation of habitat, work and leisure;
— pricing below production cost; that is, subsidies (e.g. for energy products or energy-intensive industries) lead to inefficient allocation of resources.

Energy efficiency, it is important to note, cannot be equated with economic efficiency. Thus, energy efficiency can be improved by conservation, namely by substituting energy with other input factors, with little concern for overall cost. If the non-energy inputs turn out to be relatively more costly than energy inputs, economic efficiency is not attained. Productivity growth can therefore be reached when opting for technologies that achieve overall improved efficiency, taking all factors of production into consideration, instead of being limited solely to the energy factor.[5]

Supply

Energy exists in different forms. In present technology some energy agents can be put to productive use (e.g. hydro-electric power or solar power). Other energy agents can remain untamed and unproductive (e.g. earthquakes, storms, tidal waves, etc.). Productive energy is defined here as comprising any agent that can be mastered and transformed to produce needed 'utility' in the form of mechanical work, light, variations in temperature, pressure, sound, chemical or physical changes of matter, and other functions. Few countries have been endowed with an abundant commercial energy base. Commercial sources of useful energy and their distribution among nations have, furthermore, varied considerably over time.

Commercial primary energy does not end up *in toto* in useful work. It is estimated that over 50 per cent of energy is lost in the process of production, transport or conversion. Thus, the importance of technology in the discovery, transportation, processing, storage and use of energy has become progressively greater over the years.

In the mid-1980s, over 90 per cent of the world's energy requirements were met from exhaustible resources whose formation had taken hundreds of millions of years. Chief among these depletable resources are coal, oil and gas. Over time, the mix of the energy market has undergone important changes. Wood, coal, hydro-electric power, petroleum, oil and gas have each assumed key roles at various periods and in different regions of the world. Other sources of energy (nuclear power, geothermal, synthetic crude, solar, biomass and ethanol) have so far played a moderate role (less than 10 per cent) in global energy supplies.

Factors affecting the relative cost of various sources of energy can be grouped into four main categories:

— natural or geological factors which determine the quantity, quality and ease of extraction/development of energy resources;
— technological and infrastructural factors which influence the economics of exploiting various sources of energy in different regions of the globe;
— institutional factors which mostly refer to the fiscal administrative environment and cover tax levels, incentives, distribution of risks and rewards between host country and operators with changes in the size and cost of producing energy;
— perception of political risk and the availability of the means of averting, moderating or insuring that risk.

Figures on costs are difficult to ascertain, and estimates will vary depending on the assumptions of analysts. Such estimates could, nevertheless, serve as indicators of orders of magnitude. The low cost of Middle East oil deposits can be partly inferred by reference to the large proven commercial reserves and the relatively easy extraction of oil in that part of the world (Tables 3 and 4). The economic attractiveness of these reserves will depend on their cost of production in relation to prevailing market prices. Thus, North Sea high-cost oil-fields would have remained unexploited if market prices had remained below levels necessary to insure adequate remuneration. A fall in market prices would then necessitate a reduction in fiscal charges in order to keep high-cost marginal fields in commercial production. Middle East oil producers, thanks to their prolific, low-cost fields, have been deriving substantial economic rent. This can be defined as the difference between their production cost (inclusive of normal return) and the current price in a competitive market.

The extraction costs of petroleum are influenced by a variety of factors; for example, exploration and development costs, the size of oil or gas recoverable reserves, and the rate of producing these reserves. Compared with the Middle East, the world's two largest oil producers, namely the United States and the Soviet Union, have higher levels of production as a proportion of recoverable reserves. Levels of reserves vary between ten and fifteen times current oil output in these two countries. This compares with an average forty to fifty times for major OPEC countries of the Middle East. The higher ratio in these OPEC countries is due to the combined effect of two factors: the discovery of giant prolific fields, and/or production considerably below capacity to prevent prices from falling. In terms of absolute productivity, it is worth noting that the Gulf region comes first, with an average of 3,163 barrels per day per well in 1983 (Saudi Arabia ranking highest, with 6,490). By comparison, the United States had fourteen barrels per day per well in that year.[6]

Table 3 Comparative energy costs, 1983

1983 dollars per barrel of oil equivalent on a thermal basis*	Unit costs
Middle East oil	1–5
North Sea oil	5–25
Liquids from oil sands/shale (N. America)	40–85
Indigenous coal (US)	3–10
Imported coal (NW Europe)	10–20
Indigenous coal (NW Europe)	10–35
Nuclear input break-even value†	10–25
Liquefied natural gas imports, high Btu (delivered Europe, Japan, US)	30–50
Synthetic natural gas (high Btu) from indigenous coal (US)	50–70
Liquids from coal (NW Europe)	60–100
Liquids from biomass (crops grown for fuel)	50–150
Electricity (based on conventional fossil fuel and nuclear generation)	100–150
Electricity (based on solar/wind/tidal)	100–500+

* These estimates do not include refining, storage, transmission and distribution costs to final consumers, where applicable.

† The fuel input cost required for fossil-fuelled plants to produce electricity at the same cost as nuclear stations.

Source: Shell Briefing Service, *Energy in Profile*, No. 5, London 1983, p. 8.

In the 1970s and 1980s, the discovery of oil reserves has not kept pace with past performance (Table 5). Reasons for the slower pace of net additions to oil reserves are manifold. A few major reasons, however, stand out:

— Most of the world's easy, large and cheap oil deposits seem to have already been discovered. The industry is increasingly faced with exploring and developing less accessible, moderate-sized and more expensive deposits; for example, in offshore and Arctic areas.

— Conservation of energy and the wish of consumer countries to develop alternatives to oil (coal, hydro-electricity, nuclear power, gas or solar energy) have reduced demand.

— Transnational companies have terminated their exploration–development programmes in most of the OPEC countries that have nationalised foreign concessions, although they have continued to explore in many other countries whose potential happens, however, to be less promising than that of OPEC countries.

Table 4 World oil, gas and coal reserves, 1983

Billion barrels oil equivalent	Crude oil	Natural gas	Coal
North America	83	66	1068
Caribbean and South America	34	19	72
Western Europe	16	35	431
Africa	59	41	316
Middle East	354	145	2
Far East and Australasia	20	30	324
USSR and Eastern Europe	87	232	1030
China	18	5	474
Total	671	573	3717

Table 5 Fossil fuels: reserve/production ratios, 1973–1983

	Reserve/production years		
	Crude oil	Natural gas	Coal
1973	27	37	189
1974	27	37	187
1975	30	39	195
1976	28	38	205
1977	27	38	227
1978	28	39	229
1979	28	39	231
1980	29	43	234
1981	32	49	235
1982	35	52	238
1983	35	52	240

Source (Tables 4 and 5): Shell Briefing Service, *Energy in Profile*, No. 5, London, 1984, charts 1 and 2, respectively.

— OPEC national companies have not invested sufficiently and effectively in exploration programmes, partly owing to a dearth of technical and managerial expertise or a lack of co-operation with foreign companies possessing the necessary expertise, and partly owing to policy considerations.

As far as policy considerations are concerned, a few OPEC nations which nationalised or bought out their concessionaires in the 1970s found little incentive in sustaining high levels of exploration in order to add new oil reserves or even to replace depleting reserves. In view of their already very large oil deposits (over 50 per cent of world commercial reserves are in the OPEC Middle East region),[7] a few OPEC countries deemed the size of their proven oil deposits adequate for the foreseeable future, and concluded that new oil reserves could be readily added (through extensions of existing fields, new discoveries or improved recovery methods) as and when future market conditions warranted it.

In general, the size of exploration and development budgets is in positive correlation with the level of prices and earnings—albeit with a lag. Thus, spurred by the 1979–80 oil price increases, exploration budgets rose threefold in 1981 over 1978, reaching a peak of $20.3 billion in 1981 for a group of twelve international oil companies. Exploration budgets fell thereafter with the tapering-off of open-market prices and of company earnings.[8]

In view of the fact that energy is produced from assets (oil or gas wells, coal mines, hydro-electricity plants, nuclear reactors, etc.) over long periods of time, the decision to invest does not depend solely on prevailing prices and earnings. It depends importantly on expectations of future prices and profitability. To induce increased exploration activity in petroleum, for example, expected longer-term private profitability in that activity should be at least equal to that of business subject to a similar level of risk.

Expectations of profitability are influenced by several other factors, especially:

— shifts in energy demand;
— production costs;
— excess production capacity;
— the behaviour of major energy producers, including OPEC;
— political events.

Analysts have attempted to identify measurable variables so as to build models to interpret past developments[9] and also to project possible future paths or scenarios of outcomes on the basis of certain assumptions.

Considerable caution is needed in interpreting oil and gas reserves statistics. 'Proven' reserves are those deemed to be reasonably recoverable from known reservoirs, under current technological and economic conditions (inclusive of cost–price conditions). Recoverable reserves often represent one-quarter to one-third of the original oil in place in the rock, unless enhanced recovery methods are used (the oil or gas recuperated could then conceivably rise to between 40 and 50 per cent of estimated reserves in place). Often, 'additions' to proven reserves could result from extensions to known fields, improved recovery rates or revisions to earlier estimates. Recovery rates vary considerably, depending on the properties of the deposit; they rise with the permeability of the rock, the lightness and fluidity of oil and the effectiveness of the water-drive mechanism.[10]

The energy potential of our Spaceship Earth is practically unlimited, but the development of additional commercial energy resources calls for heavy investment in know-how and capital resources. The lead periods are long in the production of significant commercial quantities of alternative sources of energy. Thus, a spokesman for the biggest energy enterprise in the world estimated in 1980 that 'a synthetic industry the size of Saudi Arabia's production [then about 9 million b/d] would cost $500,000 million [1980]—but it is technically feasible, given enough time'.[11] The time span is 'several decades', and the unit cost of synthetic oil is five to ten times the cost of producing conventional oil in the United States, and several times more than the cost of producing Middle East oil (see Table 3).[12]

Besides the commercial cost of producing, transporting, storing and using energy, the 'social' costs have to be accounted for if proper comparisons of attractiveness are to be made for different sources. These costs cover a whole array of physical and moral risks, such as potential damage to human health, the environment and public security. For example, electricity generation from nuclear 'fission' technology suffers from major actual or latent costs. These costs, albeit difficult to assess, cover:

— poor operating records and a series of small accidents;
— the disposal of nuclear waste: spent fuel-rods contain radioactive isotopes that remain potentially lethal for thousands of years;
— the perilous process of circumscribing accidents in a nuclear plant in an emergency; by comparison, electrical generating plants powered by coal, oil or natural gas are subject to less dramatic accidents and can be shut down in minutes;
— general concern over proliferation of nuclear weapons.[13]

The reliability of nuclear reactors in producing energy efficiently has fallen significantly short of expectations. The average load factor, outside the Soviet Union and East Germany, came out to 62.4 per cent between 1971 and 1980 against a targeted 80 per cent, thus representing a loss to the electricity industry of over 23,000 MW of generating capacity.[14] Large-cost overruns, long delays in completion times and changes in plant design after beginning construction have bedevilled the nuclear power generation sector. This has been particularly the case in the United States since the 1970s.[15] By the mid-1980s it became 'less costly to cancel nuclear plants that are more than 50 per cent complete than to continue building them'.[16]

Countries and companies learn, however, with experience, and one can cite the improvement in the Japanese record in designing, constructing and operating nuclear power plants. France, by standardising plant designs, has kept power costs—assuming full capacity operation—below those of coal-fired generators. Nevertheless, monopolistic practices among French suppliers of nuclear equipment have been responsible for 'excessive costs' of nuclear power.[17]

Power generation from coal has also had to shoulder additional costs to conform with official requirements. Besides stricter mining conditions to protect miners' health and safety as well as the geographical environment of mining areas, coal-burning utilities have had to install expensive 'scrubbers' in order to reduce emissions of sulphur dioxide and nitrogen oxide (e.g. to conform with the US Clear Air Act of 1970 and its amendments of 1977). Coal-burners are furthermore faced with acid rain legislation, which imposes additional high anti-pollution investment expenses.[18] With respect to energy from other sources, their significance is limited. Thus, energy from agricultural products is generally a palliative. The constraints on supplies from such a source are examined in Chapter 6.

Stocks and Excess Capacity

Variations in oil stocks could be used as an important lever in stabilising prices: drawing them down during periods of shortage and building them up during periods of excess supply. This kind of policy is usually espoused by public authorities. However, much of the oil stockpiled in the OECD area happens to belong to the commercial sector (Table 6). Company stocks fall into two categories: required stocks and discretionary stocks. The former represent minimum operating levels, imposed for 'security reasons' by the public authorities and linked to current consumption or imports, whereas discretionary stocks are decided on freely by companies.

Companies' management of discretionary stocks, as well as users' behaviour, are not conducive to countering excessive price movements. Thus, expectations of dearth and of rising prices prompt them to build inventories; on the other hand, expectations of surplus and of declining prices prompt them to liquidate inventories. It was acknowledged that the companies' stockpiling scramble of 1979–80 exacerbated price increases, while the reverse process accelerated price declines in 1982–84.[19]

Rather than building storage capacity, it can be less costly for countries

Table 6 Company and government-owned stocks in the OECD area on 1 July 1984

	Company		Government		Total	
	Mill. of metric tonnes	Days of forward consumption	Mill. of metric tonnes	Days of forward consumption	Mill. of metric tonnes	Days of forward consumption
1978	396	78	6	1	402	79
1979	385	76	22	4	406	80
1980	457	97	21	5	478	102
1981	440	99	36	8	476	107
1982	392	94	52	13	444	107
1983	355	85	63	15	418	100
1984	340	79	77	18	417	97

Note: Figures may not add, due to rounding.
Source: *OECD Observer*, Paris, September 1984, p. 28.

with oil or gas deposits to maintain productive capacity readily accessible as and when needed. However, oil-exporting countries could find it interesting to build stocks in or close to major oil-importing countries, to take ready advantage of favourable market developments. Oil stocks could also be accumulated by oil-exporting countries as a result of the need to use the associated gas produced with crude oil, notably to fuel local utilities and industries. This was 'the main reason for the existence of this [Saudi] fleet' of tankers for floating storage, according to the Saudi oil minister, Ahmad Zaki Yamani.[20]

The concept of 'excess capacity' merits elucidation. It can be considered as equivalent to that capacity readily available with little effort for sustained oil production—after taking care of maintenance and oil-fields' optimal physical production conditions, as well as the existence of appropriate downstream facilities such as storage tanks, pipelines, terminals, and/or refining–distribution capacity. A certain margin of capacity should be allocated to meet drawdown in seasonal peaks and contingencies, and cannot be considered 'excess'. Some installed capacity might also be unavailable for production, for reasons of war or conflict with neighbouring countries, blockade of transit routes, or various trade or tariff barriers.

Excess capacity, as defined above, has to be distinguishable from willingness to produce. For example, host governments could decline to produce crude oil unless they have a production use for the associated gas (instead of flaring it). Conversely, crude oil could be produced if certain industrial plants (e.g. power generation) were dependent on associated gas. Willingness to produce is further linked to economic and social costs or benefits. Thus, the rational decision to produce cannot be solely based on the assumption that marginal revenue exceeds marginal cost. There are other considerations:

- The social versus private discount rates which account for the relative urgency of obtaining revenues today or in near-term periods, as opposed to later periods. In a given period, the marginal utility of revenue to society could be zero or negative (see Chapter 6).
- Expectations of appreciation in the value of exhaustible resources—here oil—exceeding the appreciation of oil proceeds invested in other assets should lead the resource owner to hold the resource in the ground for future production, and *vice versa*.[21] Increased oil revenues accruing to a country could raise the predatory envy of interested outsiders, and the accumulation of foreign assets by an oil-exporting country could be used to pressure the country concerned.

Some analysts feel that the above considerations should be estimated and deducted from installed capacity before reaching an assessment of 'excess capacity'. Although the central concept of 'excess capacity' can be delimited in abstract with some precision, its statistical measurement is less certain.[22] Moreover, additional capacity could be brought into operation within short periods (three to twelve months) with moderate investments, thus making the concept of excess capacity 'elastic'.

With the shrinkage in demand following the post-1979–80 oil price increases, excess capacity spread to all stages of the petroleum industry—from production to downstream operations (transport, refining and distribution). Moreover, in a number of countries, marketing and refining investments made in the 1970s came on-stream in the 1980s in a contracting market. As a result, several companies encountered low profits or losses downstream and were consequently forced to close down refining and distribution facilities. Upstream, non-OPEC oil companies were protected by OPEC's production controls and price umbrella; accordingly, they managed their oilfields, outside the OPEC area, at or close to full capacity. It was the OPEC countries which supported the brunt of demand contraction to protect the world price structure of crude oil.

MARKET PRICES

Until the mid-1970s, the largest part of transnational petroleum transactions was internal to the major integrated oil companies' network, or based on long-term contracts. Shortages and surpluses did not often appear in the open market as integrated companies managed, through careful planning and logistics, to match variations in their supplies of crude oil to the effective demand for products as required to satisfy their distribution network or small operators dependent on them. Most oil-exporting countries had accepted, until the 1960s, random variations in off-takes as decided by their concessionaires, which were the major transnational companies. The nationalisation of concessions or the divestment of the vertically integrated companies in major oil-exporting countries, as well as the multiplication in the number of operators (whether independent private companies or state-owned), have contributed to the emergence of an open market in which arm's-length sales take place between unrelated companies.

The open market comprises two sub-markets: the 'term' market and the 'spot' market. The term or contract market covers sales over a forward period which could reach three years, with pre-agreed quantities for delivery and pre-agreed method(s) for calculating prices. The spot market gives price estimates of cargo-by-cargo transactions on a daily basis, or estimates of the wishes of buyers and sellers. These estimates are collected by specialised journalists and published in the trade media.

The spot market should not be seen as localised in one geographical area. It is a network of contacts between buyers, sellers and intermediaries spanning the whole world. Nevertheless, certain city ports/refining centres have had a higher concentration of operators and supporting staff; this is particularly true of the Gulf of Mexico, New York, the Arabian Gulf and Singapore, with Rotterdam assuming a pre-eminent position.[23]

Open-market prices have been particularly sensitive to excess capacity. Thus, an increase in excess capacity readily available to buyers is bound to have a depressing effect on prices. Besides supplies from new production, oil is available from accumulated inventories. The latter have been built on the initiative of companies or the authorities as insurance against unpredictable shortages of supplies or increases in demand. In the absence of these inventories with a 'compensatory function', price volatility would be higher. Indeed, open-market prices have felt the impact of distress or speculative transactions entered into by various operators. Distress-buying by refiners, distributors or consumers is prompted by the fear of running dry. Often, however, operators effect purchases, instead of drawing on abundant existing stocks (whose levels happen to exceed the normal standards of prudence), thus exacerbating further actual or expected shortfalls. Such purchases were not limited to independents, and, in 1980, majors (Shell, Exxon, BP and CFP) had reportedly 'dipped' into the spot market, paying as much as $3 per barrel above OPEC's official price.[24]

With the shrinkage in demand and the build-up of excess production capacity over the period from 1981 to 1985, price-cutting was known to be widespread among various oil-exporting countries, OPEC and non-OPEC. This was not limited to Iran or Iraq, which were eager to finance their economies because they were locked in a massively destructive war. Other countries also felt the brunt of shrinking demand and rising supplies. They had to sell below OPEC official prices in the hope of sustaining their export proceeds to cover their import bills for necessities and/or to honour their financial obligations.

The international spot market, which was negligible until the 1960s, has been rising gradually to reach the equivalent of an estimated 70 per cent of internationally transacted oil in 1985.[25] However, a cargo could be traded several times (even over fifty times) within the span of a few days.[26]

Even when, as in the first half of the 1970s, only 3–5 per cent of north-west European oil requirements were transacted by intermediaries associated with the spot market, its prices were often used by majors and independents as a basis for pricing petroleum products in some countries, such as West Germany and Switzerland.

Operators who are billed for their crude-oil supplies at the OPEC official prices obtain additional profits during periods of shortage and higher spot prices; conversely, they sustain losses in periods of excess supplies and discounts from official OPEC prices (Table 7). Prices in the spot market are typically volatile in response to the changes in the balance of demand and supply as well as to changes in expectations, which have a marked effect. As of the mid-1980s, increasingly, contracts provide a volume commitment over a period of time with quantities delivered at the spot prices of the day.

Besides non-integrated national and independent companies which participate actively in the spot market, major oil companies have also been trading for crude and products in that market, with two main objectives:

first, to meet occasional supply shortages or similarly to dispose of occasional excess production which they cannot distribute directly within their own distribution network, and second, to derive profits from strictly trading activities (e.g. by selling some of their own oil supplies at a higher price than the one they have to pay to buy equivalent quantities of the same kind). The latter trading function was resorted to, for example, by two major French groups—Total-CFP and Elf-Aquitaine.[27] These two companies, furthermore, negotiated their 1983–85 oil contracts with Iraq on the basis of a formula combining the official OPEC price and the daily spot price, but giving an increasingly greater weight to the latter—thus leading in effect to a significant disguised price discount by Iraq ($1–$2 per barrel).[28]

Table 7 Arabian Light crude oil prices, 1973–1985

Quarter	1973	1974	1975	1976	1977	1978	1979	1980	1981	1982	1983	1984	1985
Average government selling price (US$/barrel)													
I	2.41	10.95	10.46	11.51	12.09	12.70	13.34	26.00	32.00	34.00	31.00	29.00	28.33
II	2.61	10.98	10.46	11.51	12.09	12.70	15.70	28.00	32.00	34.00	29.00	29.00	28.00
III	2.82	11.05	10.46	11.51	12.70	12.70	18.00	29.33	32.00	34.00	29.00	29.00	
IV	4.41	10.59	11.51	11.51	12.70	12.70	22.00	31.33	34.00	34.00	29.00	29.00	
Average spot-market price (US$/barrel)													
I	2.10	9.90	10.45	11.50	12.20	12.70	19.65	36.10	37.65	31.00	29.30	28.55	27.78
II	2.30	10.50	10.40	11.50	12.10	12.70	27.95	35.40	33.85	32.30	28.65	28.35	27.16
III	2.60	10.00	10.45	11.60	12.70	12.80	33.60	32.85	31.95	32.00	28.90	27.65	
IV	4.50	10.45	11.50	11.90	12.70	13.80	38.35	38.40	33.75	31.75	28.35	27.74	

Sources: For 1973 until the third quarter of 1984, Shell Briefing Service, *Energy in Profile*, No. 5,London 1984, chart 9. Remaining figures from *Platt's Oilgram Price Report* and *OPEC Bulletin*, 1985.

It was reported that in the 1979 period of a seller's market, the state-owned British National Oil Corporation (BNOC) 'prefers to export 50 per cent of its production, at *top prices*, to the free markets in Germany, Scandinavia and the United States, the needs of the United Kingdom being covered by the import of crude oil purchased at *OPEC prices* from other producing countries'.[29] Indeed, in 1979, BNOC was the first to 'break' the OPEC selling price of $23.50, and rationalised a $2-per-barrel premium above world term-prices on the grounds of the 'greater security of British oil and the high spot prices'.[30] In the buyer's market period of 1983–84, BNOC sold its oil at up to a $2-per-barrel discount below the official price.[31]

BNOC's exports of a large portion of its North Sea entitlements should also be interpreted by reference to logistics and quality. Indeed, a number of major British refineries are located on the west coast of Great Britain, and several British refineries are designed to process heavier crudes (notably those of Gulf countries) than those of the North Sea.

It is interesting to note, for the sake of comparison, the restrained behaviour of a few OPEC countries, notably Saudi Arabia, during 1979 and 1980, when spot prices were surging. As one British analyst said:

> No OPEC government was prepared to raise its official selling prices, even with surcharges, as fast as spot prices. So all ex-concessionaire companies [the majors] enjoyed the privilege of lower average crude costs than third-party competitors and customers [the 'independents']. Moreover, Saudi Arabia held crude costs for its Aramco partners[32] well below the official price level accepted by most other governments in OPEC. That offered these partner companies, in effect, a further handsome subsidy per barrel.[33]

Between 1982 and 1985 the reverse situation prevailed, with spot prices below OPEC official prices (Table 7). By agreeing to sell below spot prices in 1979–80, Saudi Arabia acknowledged the value of a long-term relationship with the world's major oil groups (namely, in terms of security of outlets). By agreeing to buy Saudi oil in 1981–82 at above spot prices, the majors acknowledged the security value of long-term arrangements for supplies from a very large producer with vast commercial oil reserves.

Information on the spot or term markets is not complete. Indeed, this information depends on the willingness of traders to disclose data to the specialised media (such as *Platt's Oilgram, Oil Buyers' Guide, Petroleum Argus*). By comparison, the oil futures market provides immediate transparency. Trading in such markets is transacted in a given building, and deals are struck by means of open outcry. Once a deal has been agreed, quantity, quality, price, delivery point, date of delivery and other terms are communicated to the public outside the market by visual display, by the trade press or via electronic transmission methods. Two different groups use the futures market: producers, users or traders who want to hedge by counteracting the adverse effects of shifting prices; and speculators who want to exploit their judgement as to the way the market will move.[34] Futures markets for oil products are found in Chicago, New York and London. Although little oil is actually delivered as a result of futures (3 per cent in 1985),[35] their price movements are used by spot traders as an indicator of price expectations.

With the divestment by transnational companies of their producing concessions (often at the demand of host governments, whether by negotiation or by legislation) and of some of their downstream facilities, the number of non-integrated 'independent' actors in the oil industry has considerably increased. Independent companies, often driven by a keen rivalry, readily translate short-term changes in market conditions into price fluctuations. Beyond the real changes in the balance of supply and demand (i.e. those prompted by changes in oil's physical transactions), the impact of speculative transactions (notably in the futures market) contributes to the greater volatility of prices in the spot market.

The growth in the number of independents has thus enhanced the volatility of oil prices. In times of shortage, independents nervously bid up prices to obtain the requisite supplies. Their vulnerability and their relatively low negotiating power *vis-à-vis* major suppliers become translated into higher prices in a tight market. In an excess-capacity situation, independents are tempted to shave prices to expand sales. But they have to beware. Indeed, a major company can afford to enter into a 'price war' in a particular segment market with a view to expanding its sales against smaller competitors; it can then compensate for local losses in that segment with profits obtained elsewhere in the national market, or beyond. Price wars have raged in recent years in various market economies, particularly in the United States and Europe.[36] In one particular country, a major oil company—Mobil—provoked a price crash, ending in the sale of 'free' gasoline.[37] However, that episode was quite exceptional and remained localised.

Price wars often follow on surplus productive capacity created in response to over-optimistic expectations of growth in demand and profits. Excess capacity normally tempts producers (e.g. crude-oil producers and refiners) to expand their sales by shaving prices, given their moderate or even negligible marginal costs.

CORPORATE PRICE STRATEGIES

Building an image of reliability, of ready service and of quality through brand names creates a differentiation that encapsulates in the minds of consumers the company's efficiency and the quality of its service. In their search for expanding or protecting market niches, oil operators have resorted to such product differentiation, but with a moderate measure of success. Price remains the key factor in influencing sales. To avert the dangers of cut-throat competition, operators resorted to price *entente*, occasionally verging on price fixing. This approach has been intermittently used by both governments and private enterprises.

Economic agents, prompted by higher and stable benefits, have often sought to exploit market opportunities through group action, most notably in the areas of pricing, production, know-how and sales. The energy industry, along with other business sectors, abounds in cases of group action. One famous illustration in the period prior to the Second World War was the secret agreement, effective in the 1930s, among major transnational companies, dubbed the 'As-Is' scheme. Its details were later revealed, thanks to US Government investigations.[38] The focus of the following analysis is, however, on events since the Second World War. Since then, three major inter-governmental mechanisms have been set up by producers or consumers of energy, namely: in 1954, the European Coal and Steel Community; in 1960, the Organisation of Petroleum Exporting Countries (OPEC); and in 1974, the International Energy Agency (IEA).

Over the period 1945–70, major oil companies sought to expand sales by the competitive pricing of Middle East oil in relation to higher-cost oil. This pricing policy was prompted by the majors' large discoveries of oil reserves in the Middle East, whose potential production would outlast their concession periods. Thus, the decline in the real price of oil over the 1950s and 1960s caused considerable expansion in oil consumption. As the chief executive of a transnational oil company said, 'cheap oil' contributed to the high economic growth of the United States, as well as to the neglect of oil conservation:

> If you really want to condemn the oil industry, then I think you ought to condemn it for the simple fact that we've produced very cheap energy for a very long period of time. Now, that cheap energy made possible in many respects the tremendous growth of this country [the United States], particularly since the end of World War II. But it also made possible the growth of great big cars and the fact that when home heating oil costs 15¢ a gallon you didn't care very much about spending to insulate the house, because the cost of the heat wasn't that important.[39]

In this connection, it is worth noting that, between 1945 and 1970, major companies with large oil reserves in the Middle East concessions favoured rapid depletion (or put technically, they adopted a high discount rate for the production plans of their giant oil-fields). In the perception of these companies, wealth-maximisation strategy called for an acceleration of oil extraction as the future held risks of escalating fiscal charges and/or expropriation.

For OPEC countries' leaders, on the other hand, the 'cheap energy' policy adopted by oil companies in the 1950s and 1960s forced these countries either to 'sacrifice their own development or halt it, in order to permit or finance the development of the industrialised countries'.[40] However, it is important to note that in this period several oil-exporting countries in the Middle East were pushing their concessionaires hard to expand sales, market shares and current revenues. Thus, those Price 'sacrifices' owing to 'cheap energy' were compensated by an expanded volume. Moreover, the so-called price sacrifice was sharply reversed in the 1970s, at the expense of growth in oil demand. If price increases had been moderate and ordered over a large number of years, the adverse impact on demand would not have been as large as it has in fact turned out to be in the 1980s.

The major concessionaire oil companies, which had been negotiating with individual OPEC countries or jointly with several of them on royalty and price matters in the 1960s and early 1970s, traditionally had an important role in deciding prices and volumes of crude-oil exports. They procrastinated in acceding to host countries' demands for higher oil export prices, at a time when spot prices were surging (reaching, by 1 October 1973, a high of $10 per barrel for Arabian Light crude on the Rotterdam market).[41] These company delaying tactics in agreeing to raise their posted prices explain to a significant degree why a few OPEC countries had sought to set their oil export prices independently of their concessionaire companies. An opportunity arose with the shortage of international oil supplies created during and following the October 1973 Middle East hostilities. The Gulf countries, outside the framework of OPEC, then decided to appropriate the right 'unilaterally' to set the price of their oil exports. They raised official prices of the marker crude (Arabian Light) from $2.70 per barrel to $8.32 between 15 October 1973 and 1 January 1974 (see Table 13 in Chapter 3). The Shah of Iran then assumed the leadership role, strong in the knowledge that Iranian-auctioned oil yielded as much as $17 a barrel. The Shah's Government was then under heavy pressure to meet claims by its foreign creditors, as well as to underwrite an ambitious defence programme and grandiose development projects.[42]

Despite hardships imposed on consumers resulting from the telescoping of oil price increases between 1973 and 1974, and viewed in a longer-term perspective, these increases are only 'arbitrary in appearance', as a European official put it.[43] In fact, they were the logical consequence of a rapidly rising demand (7.5 per cent per annum over 1963–73) and a fast depletion of lower-cost oil deposits. OPEC production from January 1970 to September 1973 grew at an annual compound rate of almost 11 per cent.

The oil price increases of 1973–74 enticed operators to open up or to further develop non-OPEC higher-cost sources of energy. These included oil deposits (notably in North America and the North Sea), nuclear power and coal. The oil price increases of 1979–80, however, reached much above the replacement cost of oil and consequently provided powerful incentives to operators and governments to share in the rent windfall by developing energy sources outside the OPEC area. Concomitantly, these price increases dampened demand and acted as a boost to the conservation movement. With the increase in world oil production capacity, a price war was averted, thanks to the offsetting action of a number of OPEC countries, especially those of the Gulf, which considerably restricted their supply and maintained excess capacity.

FACTORS OF PRICE INTERPRETATION

Measuring the market value of oil poses certain problems, chief among which are:

— ascertaining whether published figures on prices are those of arm's length transactions made between unrelated operators acting independently, and whether these prices are representative of market valuation of the bulk of internationally traded oil (including countertrade transactions);

— choosing a measuring rod which can adjust current prices for inflation in order to produce comparable time-series of real prices; that is, prices expressed in constant 'purchasing power' monetary units.

In an industry which has been known for its vertical integration, concentration and term contracts, it is difficult to measure changes in market forces by sole reference to changes in official prices, published by governments or by companies. Official prices move less frequently than spot prices. The latter often react instantaneously to data which is often incomplete, partial or even erroneous. Factors influencing spot prices include, according to one trade source, 'sense and nonsense with equal indiscrimination'.[44]

In the absence of reliable data, historians of the oil industry should be careful in the interpretation of official or spot prices. There is first the question of their relevance to actual transactions: it is difficult to gauge the quantity of trade applying to reported prices. Moreover, official listing may be subject to known discounts or premiums, or to disguised ones. Undeclared discounts could thus appear in the form of favourable credit conditions, absorption of freight or insurance costs in export contracts, concessional tied aid, and a mix of oil-processing contracts for higher-quality oil, hiding discounts in the form of lower product prices.

Known countertrade deals in OPEC crude-oil exports were estimated in 1985 at about 15 per cent of the total.[45] This is often a way of disposing of oil in order to obtain the required goods and services. However, these goods and services can prove highly priced or of low quality. This is the equivalent of conceding undeclared price discounts on petroleum exports. Most oil- or gas-exporting developing countries have resorted to countertrade deals, with petroleum being exchanged for categories of diverse commodities; for example, vehicles (ships, planes, cars, rail, etc.), equipment (industrial or power generation), military wares, construction works, foodstuffs, textiles, basic materials (e.g. steel), and so on.[46]

A variant of countertrade is parallel contracts. In this case, four parties could be involved in two separate, but simultaneous and tied contracts: on the one hand, the exporter and the importer of oil, and on the other hand, the exporter of goods from the oil-importing country and an appropriate importer of such goods from the oil-exporting country. Both contracts would then use an agreed unit of account for the valuation of exchanged goods, and would aim at 'balancing' the value of goods exchanged.

Moreover, adjusting current oil prices for inflation rests on the choice of an optimal index. This index should represent, from the angle of the seller or buyer of oil, the evolution of the price level of goods and services that they normally use or consume, properly weighted. Here, the analyst is faced with various indices, each with certain advantages and limitations. Accordingly, comparing the nominal value of oil prices over a long period of time could be an important source of misinterpretation. The currency used in oil's international trade, namely the US dollar, has been subjected to

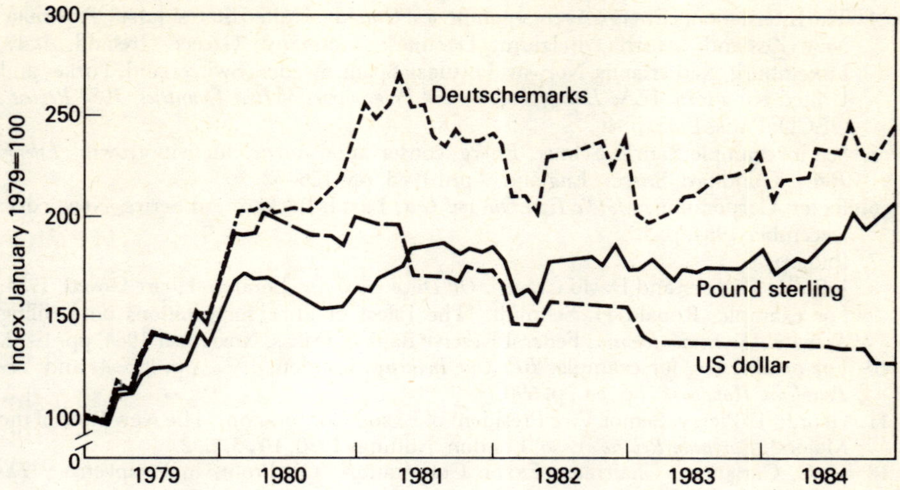

Figure 2. Brent Blend term-price changes in real terms, 1979–1984.
Source: the British National Oil Corporation, *Report and Accounts for 1984*, London, 29 May 1985, p. 6.

important changes in purchasing power. Thus, the nominal decline in the price of oil between 1979 and 1984 in terms of the US dollar was more than compensated for by a considerable appreciation of the US dollar *vis-à-vis* other major currencies, as illustrated in Figure 2.

This argument applies equally when interpreting oil revenues. Thus, the drastic decline in oil exporters' revenues in nominal dollars over 1981–84 will appear less substantial if measured in terms of constant purchasing power monetary units.

While natural endowment, the industry's structure and corporate strategies—as has been shown in this chapter—represent important factors in the evolution of world oil demand and prices, other factors have, over certain periods, been crucially significant. Among these factors are the policies and actions of governments. These are examined in the following chapters.

NOTES

1. See, for example, the World Bank, *The Energy Transition in Developing Countries*, Washington DC, 1983, p. 15.
2. One has to note that a percentage change in crude oil is not passed on directly and equally to all oil products, owing principally to the role of oil-product taxation.
3. Joseph Chai, University of Hong Kong's Department of Economics, 'China's foreign economic policy for the period 1981–1985', *China Briefing Supplement*, The Hong Kong and Shanghai Banking Corporation, July 1982, No. 9.

4. The International Energy Agency comprises: Canada, United States, Japan, Australia, New Zealand, Austria, Belgium, Denmark, Germany, Greece, Ireland, Italy, Luxemburg, Netherlands, Norway, Portugal, Spain, Sweden, Switzerland, Turkey and United Kingdom. IEA, *Energy Policies and Programmes of IEA Countries, 1984 Review*, OECD, Paris, 1985, p. 46.
5. See, for example, Sam H. Schurr, 'Energy conservation and productivity growth', *Energy Policy*, Guildford, Surrey, England, April 1985, pp. 126–32.
6. Exxon Corporation, *Middle East Oil and Gas*, Exxon Background Series, New York, December 1984, p. 5.
7. Ibid., p. 3.
8. John R. Toalster and David Craven, *Oil Discovery Costs*, London, Hoare Govett, 1984.
9. For example, Ronald H. Schmidt, 'The Effect of Price Expectations on Drilling Activity', *Economic Review*, Federal Reserve Bank of Dallas, November 1984, pp. 1–12.
10. For details, see, for example, *BP: Our Industry*, London, 1977, pp. 133–4, and *The Petroleum Handbook*, op. cit., p. 690.
11. George T. Piercy, Senior Vice President of Exxon Corporation, 'The New Role of the Majors', *European Review*, Esso, London, Autumn 1980, **10**, 3, p. 2.
12. C. C. Garvin jr., Chairman, Exxon Corporation, 'Oil Profits in Perspective', *The Lamp*, Fall 1980, New York, p. 1.
13. See, for example, Christopher Flavin, 'The Demise of Nuclear Power: What Comes Next?', *Challenge*, New York, July/August 1984, pp. 38–45.
14. David Fishlock, 'Where nuclear power stations break down', *Financial Times*, London, 29 September 1982, p. 11.
15. James Cook, 'Nuclear Follies', *Forbes*, New York, 11 February 1985, pp. 82–100; Terry Dodsworth, 'US nuclear plant problems pile up', *Financial Times*, London, 25 April 1984, p. 24; and A. S. Gurwitz and D. E. Chall, 'Nuclear Power Plant Construction: Paying the Bill', Federal Reserve Bank of New York, *Quarterly Review*, Summer 1984, pp. 48–57.
16. 'Nuclear Energy: Future Still in Doubt', *Energy*, Special Issue 1985, **10**, 2, Stamford, Conn., p. 16.
17. 'Un rapport de la Commission de la concurrence dénonce les coûts excessifs du nucléaire', *Le Monde*, 13 January 1983, p. 31; also Véronique Maurus, 'L'avenir est aux microcentrales électriques', *Le Monde*, Paris, 25 December 1984, p. 18.
18. See, for example, Michael J. Code, 'Energy: Ten Years After', *Resources*, Resources for the Future, Washington DC, Winter 1984, p. 2.
19. See 'The crucial role of oil stocks', *Petroleum Economist*, London, February 1985, pp. 42–3; see also Edward N. Krapels, 'Strategic Stocks—the balance of vulnerability', *Petroleum Economist*, September 1984, pp. 341–3.
20. Cited in *Middle East Economic Survey (MEES)*, 3 June 1985, p. A5.
21. Harold Hotelling, 'The Economics of Exhaustible Resources', *Journal of Political Economy*, **39**, April 1931, pp. 137–75.
22. See Paul Stevens, 'A Survey of Structural Change in the International Oil Industry 1945–1984', in *The Changing Structure of the World Oil Industry* (ed.), David Hawdon, London, Croom Helm, 1985, pp. 18–51.
23. See, for example, J. Bond, 'Mythe ou Réalité: Le Marché Pétrolier de Rotterdam', *Actualités*, Total-CFP, Paris, June 1985, pp. 8–10; Joe Roeber, 'Rotterdam Oil Market—what determines crude oil prices?', *Petroleum Economist*, London, May 1982, pp. 181–3; and Joe Roeber, 'Term or Spot: The Search for Stability' in *MEES*, 18 March 1985, pp. D1–D10.
24. 'The price of war', *The Economist*, 15 November 1980.
25. Ahmad Zaki Yamani, 'Serving OPEC Member Countries' best interests', *OPEC Bulletin*, Vienna, July–August 1985, p. 14.
26. Shell Briefing Service, *Trading Oil*, London, 1984.

27. Françoise Hubscher, 'The upheavals in the oil market and the strategy of Western oil companies *vis-à-vis* OPEC', *Arab Oil and Gas*, Paris, 16 July 1984, p. 30.
28. 'Elf et Total ont signé un nouveau contrat de fourniture du pétrole avec l'Irak', *Le Monde*, 19–20 August 1984, p. 14.
29. See, for example, *Road Transport and Oil*, International Road Transport Union, Geneva, July 1979, pp. 10–11 (their italics).
30. See *World Financial Market*, Morgan Guaranty Trust Company of New York, October 1979, p. 7.
31. 'Abolition of BNOC', *Petroleum Economist*, London, April 1984, p. 114.
32. They are Exxon, Texaco, Chevron and Mobil.
33. J. E. Hartshorn, 'From Multi-National to National Oil: The Structural Change', *MEES Supplement*, Nicosia, 28 April 1980, p. 3.
34. See Shell Briefing Service, *Trading Oil*, London, 1984.
35. *Oil and Energy Trends*, London, 17 May 1985, p. 3.
36. See, for example, 'Independent oil refiners are still in shock', *Business Week*, New York, 23 April 1984, pp. 79–80.
37. See 'La guerre des prix du carburant en Belgique—Une station-service bruxelloise offre de l'essence "gratuite"', *Le Monde*, Paris, 18 October 1983, p. 44.
38. See, notably, US Federal Trade Commission Staff Report, *The International Petroleum Cartel*, Washington DC, 1952.
39. Rawleigh Warner jr., Chairman of Mobil Oil Corp. 'Oil strategies, bad and good', in *Financier*, New York, September 1979, p. 32.
40. Luis Herrera Campins, President of Venezuela, Speech at OPEC Secretariat, *OPEC Bulletin Supplement*, Vienna, 24–31 March 1980, p. iv.
41. George Horwich and David Leo Weimer, *Oil Price Shocks, Market Response, and Contingency Planning*, American Enterprise Institute, Washington DC, 1984, p. 57.
42. For details, Ian Seymour, *OPEC—Instrument of Change*, London, Macmillan, 1980, pp. 113–24.
43. Robert Marjolin, formerly Vice President of the European Economic Community, 'Le Développement économique de l'Europe: la rupture de 1974 et ses prolongements dans l'avenir', *Banque*, Paris, February 1981, p. 135.
44. 'Confusion ahead for oil prices?', *Oil and Energy Trends*, London, 17 May 1985, p. 3.
45. Major deals were listed in the *Financial Times*, 7 February 1985, p. iv.
46. See A. S. Assaad, 'It is time to make barter legitimate', *Euromoney*, London, January 1984, p. 136.

3 Oil Exporters' Co-operation

GOALS AND INTERESTS

The behaviour and performance of governments and other institutions can best be interpreted by reference to their goals. Goals and their priority vary over time, depending upon each party's changing interests, and the attendant internal and external circumstances. Chief among the major goals of governments are security, wealth maximisation and political influence. These goals are bound to influence the production and pricing policies of oil-exporting countries.

Since the Second World War oil has been the main component of internationally traded energy. In the late 1940s and during the 1950s a few oil-exporting states became aware of the significance of this energy resource to their national development; they consequently began to co-operate at an informal level. Venezuela, the world's largest oil-exporter at that time, played a key role in developing these links.[1] It was not, however, until 1960 that five leading developing oil-exporting states decided to create their own permanent organisation.

Synergies

Since the inception of the oil industry in the 1860s, petroleum-exporting states have comprised a diverse group of both developed and developing countries, with a wide range of regimes and ideologies. In Baghdad in mid-September 1960 five states, all developing countries, decided to create the Organisation of Petroleum Exporting Countries (OPEC): Iran, Iraq, Kuwait, Saudi Arabia and Venezuela. Since then, eight other developing countries have joined: Qatar, Libya, Indonesia, Algeria, Nigeria, the United Arab Emirates, Ecuador and Gabon.[2]

The establishment of OPEC in September 1960 came in response to an outraged sense of injustice on the part of host governments faced with cuts made in February 1959 (about 8 per cent) and in August 1960 (about 6 per cent) in their oil export prices; these cuts directly reduced these countries' budgetary and foreign-exchange receipts. Although prompted by heightened competition from the independent companies and from Russian oil exports, the cuts were unilaterally imposed by the major concessionaire companies without prior consultation with, or warning to, host governments. These concessionaires then comprised seven major transnational companies: five American (Exxon, Texaco, Mobil, Gulf) and two European (Royal Dutch/Shell, British Petroleum).

The creation of OPEC has enabled member countries to speed up the process of wresting from the Western majors their effective control over crude-oil production in the OPEC nations. In the perception of OPEC founders, and of those who joined them later, the Organisation thus serves as a collective defence mechanism for developing countries heavily dependent on oil exports (measured in terms of foreign-exchange earnings, budgetary receipts or contribution to GNP). In particular, OPEC was conceived as a means of fending off or reversing adverse situations such as:

— pricing and/or production policies of foreign concessionaires which could undermine the socio-economic stability of individual host countries;
— excessive economic dependence *vis-à-vis* a few major industrial countries, whose economic policies could maintain oil-exporting countries in a relationship of relative underdevelopment as mere suppliers of raw materials;
— the power of industrial countries or of their major companies in playing off one oil-exporting country against another, or in applying punitive measures;
— instability of the energy market, and its negative impact on the oil-exporting countries' balance of payments and development.

OPEC members foresaw the advantages that could be derived from joint deliberations and joint actions within the framework of a collegiate vehicle. Such advantages cover the raising and sustaining of oil revenues, improved terms for converting a significant portion of their oil-sale profits into productive assets (physical, human and financial) in accordance with their national developmental goals and priorities, and larger economic and non-economic gains in regional or international affairs within a 'Third World' solidarity perspective.

The sense of vulnerability, economic and non-economic, of the developing oil-exporting countries was variously conveyed by many of their spokesmen. Thus, Ali Attiga, the secretary-general of the Organisation of Arab Petroleum Exporting Countries (OAPEC) pointedly said in 1981:

> As exporters of one commodity and importers of thousands of consumer and capital goods and services, they have only a few options. Their export and import markets are under the control of a few industrial advanced countries. Their domestic currencies and foreign assets are, in fact, dependent on the monetary and fiscal policies of even fewer countries. Their access to modern technology and know-how is highly restricted by a combination of their own internal limitations and the terms and conditions usually imposed by the technology-exporting institutions

and governments. Their strong feelings for internal security and external defence needs, whether real or exaggerated, readily induce them to spend an increasing portion of their oil revenues for purchasing suitable and unsuitable modern arms and other military hardware. Because these countries can pay for their civil and military imports in cash, they provide attractive markets for the salesmanship of the exporting countries. All these features of the oil-exporting economies reinforce their dependence on foreign markets and tend to diminish their indigenous drive for self-reliance and regional co-operation.[3]

The aggregation of petroleum-exporting countries, however, conceals wide differences among them. The national objectives and interests of individual countries reflect the philosophy and perceptions of their leadership and elites. OPEC countries cover the whole gamut of regimes, stretching from 'revolutionary' or authoritarian regimes on the one hand, to quasi-Western-style parliamentary democracies on the other. Furthermore, OPEC countries' conditions vary considerably, notably with respect to socio-economic structure, size of population, institutional development, import requirements, resource endowment, internal/external pressures and geopolitical attitudes.

Economic Divergences

The great diversity of oil-exporting countries can be assessed by reference to a number of objectively measurable criteria such as: population, area, oil reserves, oil reserves' longevity (set against production), oil revenues, imports, the 'level of development', the degree of 'dependence' on one or a number of commodities for export proceeds and the degree of 'dependence' on one or a number of trading partners for the supply of essential commodities. Certain quantifiable elements in the profiles of oil-exporting countries illustrate the wide disparities. Countries vary from the highly populous (such as Indonesia) to the thinly populated (Kuwait, Gabon, Qatar, the UAE); from those with large commercially exploitable oil reserves (e.g. Saudi Arabia and Kuwait) to those with limited reserves, whether measured in absolute terms (e.g. Gabon) or in terms relative to population (e.g. Indonesia) (Table 8).

OPEC countries have sought effective means of influencing their export profits and setting the pace of depletion of their oil reserves. The control of depletion is influenced by factors that vary in significance from one country to another. Among the more important are the size of the country's actual and potential commercial oil reserves, the competitive conditions of the energy market, and import 'needs' for development and/or security purposes. Countries with limited oil reserves and with large, pressing domestic needs, owing notably to a relatively large population with low per capita income levels and/or high security requirements, have been anxious for higher, more immediate export profits—whether through maximalist

Table 8 Oil exporters—selected data

	Area 1000 km²	H Population mid-1983 million	R Recoverable reserves million tons 1984[e]	P Production million tons 1984	R/P years	R/H tons	GNP/H[e] US$ 1983
Algeria	2,383	20.6	1,146	29.5	39	56	2,400
Ecuador	284	8.2	194	12.8	15	24	1,430
Gabon	267	0.7	70	8.0	9	100	4,250
Indonesia	1,919	155.8	1,446	70.5	16	7	560
Iran	1,648	42.5	6,554	105.0	62	154	n.a.
Iraq	435	14.7	5,933	58.5	101	404	n.a.
Kuwait	18	1.7	12,807	58.0	221	7,534	18,180
Libya	1,760	3.3	2,806	52.5	53	850	7,500
Mexico*	1,973	75.1	6,845	150.0	46	91	2,240
Nigeria	924	93.6	2,250	68.0	33	24	760
Norway*	324	4.1	1,092	34.5	32	266	13,820
Qatar	22	0.3	441	18.8	23	1,470	21,170
Saudi Arabia	2,150	10.4	25,512	235.0	109	2,453	12,180
United Arab Emirates	84	1.2	4,103	55.5	74	3,419	21,340
United Kingdom*	244	55.9	1,819	125.0	15	33	9,050
Venezuela	912	17.3	3,692	95.0	39	213	4,100

* Not member of OPEC
n.a. = Not available.
[e] Estimated.
Sources: *The World Bank Atlas 1985*, Washington DC; and Esso, *Oeldorado '84*, 1985.

price policies and/or through larger market shares or volumes. This is to be contrasted with the concern of countries with large oil reserves, which have preferred oil price stability at levels which do not encourage radical substitution for oil by consumers. In the first group of countries one could include Algeria, and in the latter group of countries, Saudi Arabia.[4]

Differences of policy exist even within a country's governing group which appears, at first sight, to be monolithic. This is the case in Saudi Arabia. As a former US ambassador to that country commented in 1978, the oil policy of:

> Crown Prince [since June 1982, King] Fahd and Minister of Oil Yamani on prices and production certainly does not have the support of most of the people in the country, or even some of the people in the Government. Most believe prices should go up and Saudi production go down.[5]

Leaders' positions, however, do change in response to new circumstances.

The Saudi ruling elite can be divided into two main streams:

- 'conservationists' advocating 'moderate' production in order to husband resources for future generations;
- 'expansionists', who feared that tight supplies and high prices could permanently drive consumers from natural crude oil into alternative fuels, thus leaving vast volumes of unwanted oil reserves in the subsoil.

The implication for Western economies of the Saudi 'expansionist' approach in 1979 and 1980 was put by a Saudi government official, as follows:

> We could hardly have been blamed for deciding that it was not in our interest to rapidly deplete our valuable oil resources or to run up current account surpluses that promptly depreciated in value. Far better to slow down our growth and limit oil exports to the amount needed to cover our immediate and more modest development needs. The consequence for the West of such a development strategy would have been unfortunate. It would have meant even more rapid increases in the price of oil than those which have taken place since the mid-1970s. It would have meant chronic supply shortages, more inflation, slower growth and higher unemployment in the West.[6]

The Saudi planning minister, Hisham Nazer, stated in April 1981 that Saudi Arabia could easily meet its financial requirements with half of its 1980–81 output (which averaged about 10 million b/d). He candidly admitted that Saudi Arabia was overproducing 'with the specific objective of forcing other producers to temper their price demands and agree to a long-term pricing strategy'.[7]

When it comes to identifying the countries that command greater influence within OPEC and in the market-place, the criterion of present and potential productive capacity is a key factor. Saudi Arabia in the 1970s and 1980s comes first in this respect. In the years 1979–81 Saudi oil reserves and exports represented about 38 per cent of OPEC's total, and the Saudi oil minister openly declared, in April 1981, 'We engineered the Glut' to force other OPEC countries to moderate their drive for higher prices.[8] In the depressed oil market of 1981–85, Saudi officials became concerned about arresting a sharp drop in oil prices and in their volume of exports.

Saudi Arabia may find it in her long-term interest not to oppose a gradual decline in the real prices of oil exports (from their level in 1985). Given its

Table 9 Investment strategies compared, 1971–1981 (% p.a.)

	Average real yield	Volatility: standard deviation
US dollar bonds	−4.0	6.0
US equities	−0.9	19.4
Sterling deposits	0.1	12.0
US dollar deposits	1.3	2.2
Deutschemark deposits	2.8	13.3
Yen deposits	4.1	15.3
Gold	16.5	45.2

Source: Kevin Pakenham, 'The Dollar Investor in a Floating World, 1971–81', *Amex Bank Review*, London, 24 May 1982, p. 7.

huge petroleum reserves, Saudi Arabia's interest lies in discouraging the development of alternative sources of energy, and/or further developments in conservation and oil efficiency. With declining oil prices (as was the case between 1981 and 1985), shifting from a strategy of oil in the ground to one of building up domestic or foreign assets would appear to be profitable. This was not the case in the 1970s (Table 9) when yields on external financial assets (dollar and non-dollar) were negative or moderate, while oil was appreciating.[9]

Saudi officials, however, do not look solely at real rates of return on official foreign financial assets versus oil in the ground. They are anxious about generating adequate cash flows to disburse in the domestic economy. The impact of a drastic cut in government expenditure could be harmful to the private sector and could be a source of social upheaval. The cut in government expenditure reached 34.2 per cent over the period 1981–85 (Table 10).

Table 10 Saudi Arabian budget, 1978–1985

Fiscal years†	Revenue*		Expenditure*		Balance $ bn‡
	$ bn‡	% change on year	$ bn‡	% change on year	
1978	38.38	+3.7	43.52	+11.3	−5.14
1979	62.86	+63.8	56.06	+28.8	+6.80
1980	104.53	+66.3	71.04	+26.7	+33.49
1981	108.88	+4.2	84.22	+18.6	+24.66
1982	71.77	−34.1	71.40	−15.2	+0.37
1983	55.29	−23.0	64.69	−9.4	−9.40
1984	47.41	−14.3	60.48	−6.5	−13.07
1985	55.40	+16.9	55.40	−8.4	0

* Actual expenditures and revenues for 1978–84, Budget data for 1985.
† The fiscal year begins on 22 March.
‡ Saudi Riyal data converted at average year exchange rates. 1985 figures converted at $1/SR3.61.
Source: Ministry of Finance and National Economy of Saudi Arabia; cited by the *Amex Bank Review*, London, 28 May 1985, p. 4.

Excess capacity in OPEC countries in the 1980s, combined with deficits in their balance of payments (Table 11) and high real interest rates, are certainly factors which favour competition among the OPEC debtor countries and the maintenance of 'soft' prices. Rivalry between the two warring members of OPEC has also contributed to competitive prices. Thus, the Iraqi oil minister Qasim Ahmad Taqi, declared in May 1985 that one of the objectives of their attacks in 1984 and 1985 on Iran-bound oil-tankers, as well as on Iranian oil terminals, was to 'cut down Iran's oil

Table 11 Payments ($ bn) and production data (000 barrels per day)

	Current account		March 1983 quotas	Nov. 1984 quotas	Mid-1985 sustainable capacity
	1983	1984			
Algeria	−0.1	—	725	663	800
Ecuador	−0.1	−0.2	200	183	300
Gabon	0.1	0.4	150	137	200
Indonesia	−6.3	−4.0	1,300	1,189	1,700
Iran	0.1	—	2,400	2,300	3,000*
Iraq	−3.4	−4.0	1,200	1,200	1,300*
Kuwait†	4.6	7.0	1,050	900	2,300
Libya	−1.7	—	1,100	990	1,500
Nigeria	−4.7	−1.0	1,300	1,300	1,800
Qatar	2.0		300	280	500
Saudi Arabia†	−18.4	−12.0	5,000	4,353	8,800
UAE	3.5	3.0	1,100	950	2,125
Venezuela	3.7	4.0	1,675	1,555	2,500
Total	−20.7	−6.8	17,500	16,000	26,825

* Excluding shut-in capacity caused by war.
† Including share in neutral zone.
Sources: For quotas, OPEC Secretariat; for sustainable capacity, estimates by *Petroleum Intelligence Weekly*, New York, 20 May 1985, p. 7; for current account balances, Shell Briefing Service, *Oil, Gas and Coal in 1984*, No. 2, p. 6.

exports' and to force Iran 'to offer discounts on its oil sales' to entice buyers to brave Iraqi missiles. The end result sought by Iraq was attained, namely 'a significant reduction in Iran's oil revenue' during the period of hostilities.[10]

The Saudi–US Relationship

OPEC's largest oil producer, Saudi Arabia, explicitly links its production policy to political security as well as to economic interests and goals. The Saudi Government looks to the United States for the protection of the monarchy, to bring lasting peace to the Middle East and to assist in Saudi Arabia's internal development plans.[11] The two countries happen to be major trading partners, and their sense of interdependence has been further heightened by the oil-import requirements of the United States and the substantial Saudi assets held in US currency. In 1981 President Reagan pledged US support to the Saudi monarchy against external as well as internal threats.[12]

It is well known that Saudi Arabia, among OPEC countries, has been more responsive to American wishes in pricing or determining the size of its oil exports. Saudi rulers and several officials view their bilateral relations with the United States as an almost 'perfect' model of co-operation in the political and economic field. The only stumbling block, as put by the Saudi oil minister, Ahmad Zaki Yamani, is the lack of US support for the right to self-determination of the Palestinian people. He ranks the 'threat' arising out of this problem above that of potential Soviet incursions or subversions. In 1981 he said:

> To the Saudis, there are only two threats in this world: international communism and Israel. The first helps to reinforce our friendship with America, while the second is a threat to that friendship. It is difficult to conceive of a compromise in a situation like this. The second is far more tangible and more in evidence than the first one, and an actual danger is obviously worse than a potential danger.
>
> The educated among the Saudis are fully alive to the fact that Israel is the opening through which the Russians slip into the Arab world, that it is the communists' entry visa to our countries. And the more support the US gives to Israel, the better is the climate for the Russians in the region. This is fully borne out by the pattern of Arab–Russian relations during the last three decades. Given this, we in Saudi Arabia find it very strange to hear that the communist threat is the justification for the special relationship between the US and Israel. We believe that it is in the interest of the United States and the West that peace and stability should reign in the region. We also believe that Russia stands to gain from the present situation and that Israeli policies, particularly those of the current government, are at one with Russian policies. When the US, due to internal political pressures, refrains from performing its duty in bringing peace to the area, it thereby serves the Russian interests, to the detriment of its own interest and perhaps those of Israel itself.[13]

Saudi Arabia depends heavily on US technical assistance in the military and non-military field. As of June 1982, 1,500 federal employees were working in Saudi Arabia or in support positions in the United States and their salaries and benefits were being reimbursed by the Saudi Arabian government. By December 1981 these costs had reached a total of $507 million, running at $80 million annually and covering three programmes: the United States–Saudi Arabian Joint Commission on Economic Co-operation, the Geological Survey and the US Army Corps of Engineers. Through the United States–Saudi Arabian Joint Commission on Economic Co-operation, the Saudis deposited over $1.1 billion between 1975 and 1983, of which some $800 million were disbursed in the US market.[14] The total value of the construction work for military facilities in Saudi Arabia amounted to about $16.3 billion by the end of 1981.[15]

US official involvement in the Saudi scene can be ascertained by reference to the advantages obtained by the government from five aircraft with Airborne Warnings and Control Systems (AWACS). The extra advantages to the US government are listed in Table 12. In 1981 the Saudis

Table 12 US terms and conditions of sale of AWACS to Saudi Arabia

Standard agreement		Additions
Use of AWACS:	Defensive use only	No flights outside borders (without prior US consent)
Security of technology:	Protect classified equipment with procedures similar to US procedures	US government approval of security plan US government inspections High technology security facilities Only US and Saudi personnel have access to equipment and documentation New information security agreement Computer software (machine language) remains US government property
Third-country transfers:	No transfer of equipment without US government approval	Third-country modifications to equipment forbidden Third-country personnel forbidden to perform maintenance
Information-sharing:	None	AWACS data exchanged between United States and Saudi Arabia at all times No AWACS data to other countries without prior and mutual consent of United States and Saudi Arabia

Source: Secretary of state Alexander Haig, 'Dangerous Illusions and Real Choices on AWACS", *Current Policy*, No. 324, Washington DC, 5 October 1981, p. 2.

agreed to these conditions and to paying $8.5 billion in cash. One former US ambassador to Saudi Arabia commented on this transaction thus:

> the real question is whether the United States will give all of the information that the AWACS gather to Saudi Arabia, not whether the United States will have access to it. The sale would be extraordinarily advantageous to the United States; never in our history have we had a base (and a base is precisely what the five AWACS would be) flying above another country in a sensitive area and paid for by that country, with a huge monetary and strategic profit accruing to the United States.[16]

In 1983 and 1984 the Saudi Government tried to derive some economic leverage from military contracts with the United States. It sought investments by US manufacturing enterprises in selected joint Saudi ventures for sums up to 35 per cent of the contract value. These would be expected to manufacture products which could compete in the international market both in terms of price and quality. The US partner should provide advanced technology and commit itself to at least 40 per cent of the joint venture's equity in the first five years and 30 per cent of equity in the next five years. In return, the joint venture would enjoy a ten-year tax holiday, concessionary credit for 50 per cent of the project cost, subsidised water and electricity rates, a commitment to give them preference in the procurement of supplies to the Saudi public sector and also possibly official export support.[17]

While allied to, and dependent upon, the United States for its security (as evidenced, *inter alia*, by the presence in Saudi Arabia of US military advisers—estimated at some 4,000 in 1981), the Saudi Government does not want to show undue dependence—lest it offend domestic nationalist feelings or regional sensitivities. This concern is shared by other smaller Gulf states.[18] Relations with France are also known to be close. Indeed, it was France which supplied officers and equipment to assist Saudi royal troops in November 1979 in storming Mecca's Grand Mosque, which had been taken over by opponents of the Saudi monarchy.[19]

US Government agencies (the Departments of Agriculture, Commerce, Energy, Interior, Labor, Treasury and Transportation, as well as the Farm Credit Administration, General Services Administration and National Science Foundation) act as agents for Saudi Government departments. This relationship, according to an official US source, brings 'significant trade and political benefits to the United States'.[20]

THE NATURE AND FUNCTIONS OF OPEC

To understand the nature of OPEC we may begin by assessing the extent to which it fits any of the models attributed to it by various writers. The paradigms most often used are those of a political alliance, a bargaining agency or a cartel.

Political Alliance

Let us first consider the hypothesis that OPEC is a political alliance. The fact is that ideological and political differences, border disputes and other sources of conflict have existed among OPEC countries since the inception of the Organisation. To illustrate, in 1960 and 1961, Iraq laid claim to the entire territory of Kuwait. Iran and Iraq have fuelled disturbances in each other's land, finally leading, in 1980, to the outbreak of large-scale military hostilities. Diplomatic relations have often been suspended between fellow OPEC members: these include, Iraq–Kuwait, Iraq–Iran, Libya–Iraq, Saudi Arabia–Libya, Libya–Iran.

The Iran–Iraq war is one of the most dramatic conflicts between any two countries since the Second World War in terms of:

— the size and status of the countries;
— the scope and duration of hostilities;
— the massive death and destruction inflicted;
— the impact on regional and international events.

It is difficult to ascertain responsibility in the initiation of hostilities. Independent observers note, however, that Iraq was to some extent provoked. Indeed, the revolutionary Islamic regime in Iran which seized power from the Shah in February 1979 appealed to Shiah communities on the Arab side of the Gulf to overthrow their Sunni rulers.[21] The Iraqi Shiah had genuine grievances, chief amongst which was their very small share of power. These grievances were exploited by the Islamic regime in Iran, which supported (by both propaganda and material resources) insurrectionary attempts by local disgruntled Shiahs led by their religious leader Imam Muhammad Bakr Al-Sadr, and grouped under the Al-Dawa movement. It is important to note that the spiritual and temporal leader of the Iranian Islamic revolution, Ayatollah Khomeini, felt profound animosity towards the Iraqi president, Saddam Hussein. The latter, at the request of the Shah, asked Aytollah Khomeini to leave Iraq in 1978, where he had spent fourteen years in exile.

The war began in February 1979—according to Iraq's leaders—with the establishment of the new Islamic regime in Iran, which Iraq accused of a series of border violations, notably the bombardment of Iraqi villages and towns. They further alleged Iranian support for Al-Dawa's attempt on the life of Tariq Aziz (the third-ranking leader of the Iraqi regime) on 1 April 1980. Iran's shelling of the Iraqi border area of Khanekin and Mandali on 4 and 5 September 1980 provoked Iraq's leadership into launching a massive counter-attack. One informed analyst commented, 'Iraq began large-scale, formal hostilities, but it was Iran that began the war'.[22]

The Iraqi army crossed the international frontier on 22 September 1980. Among the myriad reasons for taking this momentous decision, three seem particularly significant:

- Iraq's leaders wished to regain territorial sovereignty over the Shatt-al-Arab waterway, thus abrogating the Iran–Iraq treaty signed in 1975 in Algiers, which Iraq claimed it had signed under duress.
- Iraq's secular Ba'ath party sought to protect its regime by military means against Imam Khomeini's 'export' of Iran's Islamic revolution to Iraq's Shiah population, which constitutes the majority of that country's Arab-speaking people.
- Iraq's president, Saddam Hussein, aspired to assume pan-Arab leadership in defending the eastern Arab flank against non-Arab 'Iranian hegemony'.

The timing of the start of Iraq's general military campaign can be explained by reference to a number of complex factors, chief amongst which are:[23]

- Iraqi leaders' perceptions of their superior military might;
- the apparent weakness of the Iranian regime owing to internal uprisings, revolutionary purges and dissensions within the ruling class;

— the international isolation of the Iranian regime following the taking as hostage of American embassy personnel and the ensuing Western boycott;
— Iraq's expectations of wide, active support from Arab governments;
— Iraqi military expectations of support from the ethnic Arab minority living in south-west Iran;
— a general belief in a quick conclusion to hostilities.

The Iraqi leadership's reasoning proved unrealistic, however. It is an irony of history that hostilities broke out about the same time as the planned twentieth anniversary of OPEC, which had been created in Baghdad in September 1960. The celebrations were to be attended by heads of state of OPEC countries, including Iran and Iraq, in the capital of the latter (Baghdad). The then Iranian minister of oil, Mohammed Jawad Toudguyan, was shown on Iraqi television in Baghdad in November 1980, not as a welcome official guest but as a captive whose party had been taken prisoner near the beleaguered Iranian oil city of Abadan.

Beyond Iran and Iraq, long-standing border disputes between some OPEC countries and their neighbouring states have been the source of sporadic conflict; for example, between Ecuador and Peru, Venezuela and Colombia or Guyana, Libya and Malta or Tunisia, Saudi Arabia and the Yemens (North and South), Algeria and Morocco, Nigeria and Cameroon or Chad.

The fact that OPEC is not a political alliance does not preclude the existence of political alliances within and across the OPEC group. Thus, as is well known, the Arab Gulf states (notably Kuwait, Saudi Arabia, the United Arab Emirates and Qatar) stood by Iraq in the 1980s in its armed confrontation with Iran. Logistical support, ammunition, food and interest-free loans have been liberally given. Following on from Syria's blockade, since April 1982, of Iraqi oil transiting through its territory, the governments of Kuwait and Saudi Arabia further agreed to provide Iraq with crude oil in exchange agreements—estimated in 1984 at 250,000 barrels per day from the Arabian Oil Company which operates in the Saudi–Kuwaiti offshore Neutral Zone, and between 60,000 and 70,000 barrels per day of Arabian Light crude from Saudi Arabia itself. These arrangements fell within a 'time-exchange crude deal', defined as 'receiving crude now and delivering it back in kind at a later date'. This arrangement is expected to continue into 1986.[24]

Arab Gulf states were concerned at the destabilising impact on their own regimes of the Iranian Islamic revolution. In 1979 and 1980 the revolutionary eruption in Iran, spearheaded by Shiah Muslim leaders, triggered overt dissatisfaction among Shiah groups in Arab Gulf states ruled over by leaders belonging to the Sunni branch of Islam. Popular uprisings in the latter countries were kept under control by the regimes concerned. The Iran–Saudi rivalry in politics and oil has been eloquently expressed by

Iran's oil minister, Mohammad Gharazi, thus: 'Every barrel trimmed from Saudi Arabia's output and added to ours is beneficial to our revolution'.[25]

The sister organisation of OPEC, OAPEC, is also far from representing a cohesive group. The Organisation of Arab Petroleum Exporting Countries (OAPEC) was supposed to bring countries sharing the same Arab culture closer together by co-operating over their petroleum interests.[26] This goal has proved elusive, with intra-OAPEC dissensions. To illustrate, OAPEC members, between 1980 and 1985, were divided in their support for the two warring protagonists in the Gulf. In particular, Syria and Libya actively supported Iran (a non-Arab country and consequently not a member of OAPEC), while the majority of the other OAPEC members supported Iraq, with Algeria declaring its neutrality.

Syria's support for Iran went beyond the supply of strategic materials to sustain Iran's war effort to the point of blocking the passage, beginning on 10 April 1982, of Iraqi oil through pipelines crossing Syrian territory. In 1982 the capacity of these lines reached approximately 800,000 barrels per day. One estimate put the annual loss of oil revenue to Iraq at $8 billion per year.[27] The interruption of Iraq's oil supplies through Syria has occurred despite the fact that the two countries' leaderships espouse the same ideology (that of the pan-Arab Ba'ath party), though each is governed by a rival faction. Syria has reportedly obtained from Iran a gift of one million tons of crude oil annually as a reward for its pro-Iranian stand.[28]

In the midst of the Iraq–Iran hostilities, and prompted by an acute perception of omnipresent threats of intervention by the superpowers either directly or via surrogates, the foreign ministers of six Gulf countries agreed on 4 February 1981 to create permanent machinery which would enable them to co-operate in warding off potential external and internal threats. The countries concerned—Bahrain, Kuwait, Oman, Qatar, Saudi Arabia and the United Arab Emirates—formed the Gulf Co-operation Council (GCC), whose declared aim is the protection of their mutual security (external and internal) and the deepening of their economic and political co-operation.[29]

These countries share a number of important features. Not only are their economies heavily dependent on oil, but they also have a broadly similar demographic background, conservative ruling dynasties, and they all favour progress by evolution. Achieving unity is likely to be a long-winded process. As one informed analyst put it, 'a long history of fragmentation, feud, personal rivalry and in times mistrust, cannot be eliminated miraculously by the stroke of a GCC'.[30]

The Gulf countries certainly illustrate the deep-seated suspicions and fears that exist between some OPEC member countries. The leaders of the GCC are concerned about the hegemony of other Gulf countries (i.e. Iran and Iraq). Indeed, Iraqi president, Saddam Hussein, commenting on GCC's apprehensions about Iran, was reportedly quoted by the French foreign minister as having said 'Yes, but they [i.e. the GCC countries] are also very afraid of me'.[31]

Collective Bargainer or Cartel?

In the 1960s and early 1970s, the OPEC organisation was used by member countries as a vehicle for collective bargaining *vis-à-vis* transnational companies which held concessions in these countries. Group bargaining covered questions of oil export prices, fiscal matters and equity participation in domestic ventures.[32] For such activities, one scholar considered that 'except for its duration the bargaining [through OPEC] did not differ radically from industry-wide industrial union bargaining in American industry'.[33] Another writer supported the analogy of trade unions thus:

> The Group of 77 [the developing countries] in this trade union perspective resembles a General Workers' Association. One group of workers, whose services are essential to the bourgeoisie, to the functioning of the bourgeois economy and even to the stability of the bourgeois institutions, is particularly well organised, namely the workers producing oil [OPEC] for the bourgeoisie [the IEA states]. In this perspective, OPEC is a trade union and OAPEC its ethnic sub-associate club. These workers have just made a successful strike, of a revolutionary character, not only achieving a substantial increase in wages, but even changing property relations, i.e. taking over their places of work.[34]

However, the 'trade union' model does not seem to fit the functions of OPEC as an institution adequately. This model implies that its members fully entrust their union to negotiate appropriate salaries and other rewards commensurate with their labour input. Oil revenues are not, however, the result of human exertion, but represent the proceeds of a sale, in instalments, of a patrimony. Moreover, Western developed countries formed the International Energy Agency (IEA), but this cannot be regarded as an employers' association. In fact, there have been no institutional contacts between OPEC and IEA, and certainly no collective bargaining. OPEC has been more often branded a cartel. The latter, in the conventional sense, centralises key economic decisions. The powers of a cartel cover three major areas:

— Fixing the group's output and/or price. An optimal distribution of quotas among the cartel's members would call for the equalisation of marginal cost for each producing field with marginal revenue derived from aggregate demand.[35]
— Redistributing income among members of the cartel to 'compensate' or 'reward' those members who have shut down their domestic production in favour of others.
— A central mechanism to audit and enforce collective decisions. Enforcement would call for an authority to administer rewards and penalties.

The institution of an OPEC cartel has proved a challenge, owing to the fact that members are sovereign states anxious first and foremost to protect their national interests. This author was involved in 1966 (as adviser to the Kuwait delegation) in the designing of an OPEC-wide production programme with a view to shoring up prices. This scheme, and the attempts that ensued, were balked by the overt or covert rivalry of governments and companies. Collective agreement on the choice and weighting of objectively measurable variables as a basis for the fixing of quotas thus proved impossible.

One OPEC country minister conceded in 1981 that the Organisation had lacked the defined rules of a cartel. He stated, '[a] cartel is much more disciplined than a loosely-tied organisation like OPEC. In a cartel there are more definite rules of the game. I think I would describe OPEC as not much different from the Wheat Council or any other organisation of that sort'.[36] The International Wheat Council groups together the major wheat-exporting countries—nearly all developed countries. Under the leadership of the United States, the world's major wheat exporters meet informally for consultations with the aim of stabilising world wheat supplies.[37]

However, after 1981, OPEC was increasingly guided by the cartel model to protect the world oil price structure from a possible collapse. Indeed, spurred by OPEC's price increases in 1973–74 and 1979–80, non-OPEC countries were able, over the 1970s and 1980s, to increase their oil supplies at the expense of OPEC countries. The latter, with sagging world energy demand, felt the brunt of excess supplies through further reductions in their exports and/or significant reductions in market prices. In order to avert the breakdown of their general price structure, OPEC countries envisaged a 'crisis cartel' aimed at stabilising prices of crude oil exports. To achieve this objective, most of the OPEC ministers sought in March 1982 to voluntarily reduce their oil export volumes with a ceiling of 18.5 million barrels per day. However, the populous countries (notably Nigeria, Iran and Algeria) openly called for substantial cutbacks in Saudi Arabia's large oil exports. The Saudi Government finally agreed to reduce its share of exports and/or provide needy fellow members with economic assistance in periods of crisis as a means of protecting the cohesion of the group, and consequently the stability of the world oil price structure. Thus, Ahmad Zaki Yamani, the Saudi oil minister, indicated early in 1983 that his country had provided financial support to Nigeria to shore up its balance of payments and induce Nigeria to maintain OPEC's official price.[38] The credit in question amounted to $400 million and was granted in mid-1982, according to Nigeria's finance minister (then Victor Massi).[39]

With the drying-up of Saudi credit, the sharp decline in open-market prices and the state-owned British National Oil Company (BNOC) reduction in the North Sea oil official price by $3 a barrel to $30.5 a barrel, Nigeria could not support the strain on its balance of payments. In February 1983 it reduced its oil export prices, without consulting its fellow OPEC members, from $35.5 to $30 a barrel. It further declared its wish to match directly 'cent for cent' any cut in British oil prices.[40] (The two countries produce some similar qualities of crude oil.)

The OPEC Conference of Ministers, after painstaking, arduous negotiations sustained by looming fears of a price war which could lead to a spiral in price cuts with bleak consequences for their oil revenues, managed to agree on 14 March 1983 to reduce the official price for the Saudi reference crude (Arabian Light) from $34 a barrel to $29. This was the first official price cut in twenty-three years since OPEC's establishment in 1960. The price agreement was to be bolstered by a production programme allocating quotas to individual member countries, within a group ceiling of 17.5 million barrels per day.

In response to incessant Nigerian solicitations for a larger share of oil exports, the Saudi Government agreed in July 1984, within an OPEC Conference of Ministers, to reduce its own exports share in favour of Nigeria (by up to 150,000 barrels per day for the balance of that year). Nigeria had been in the throes of a financial crisis, burdened by a large foreign debt, which it was unable to honour.

With the continued erosion of oil export prices in the open market, OPEC countries had to further reduce their exportable production—effective from 1 November 1984—in order to prop up prices. The global reduction was set at 1.5 million barrels per day, bringing down the previous production ceiling of 17.5 million barrels per day to 16 million barrels per day, or by 8.57 per cent. The largest quota reduction was sustained by Saudi Arabia (647,000 b/d). Iraq's and Nigeria's quotas remained unchanged (see Table 11).

In January 1985 the single-reference-price crude (Arabian Light) was abandoned in favour of a grid of reference prices for different sources and qualities of crude, the whole forming a composite benchmark. Moreover, it was agreed to institute a monitoring system of members' production levels and prices. An independent group of auditors (Klynveld Kraayenhof of the Netherlands) was entrusted with the task of inspecting quantities of oil exports by member countries, as well as prices.

The practices of exceeding agreed production ceilings and/or shaving official prices have been widespread among OPEC members. For its part, Iran attributed its 1980–85 policy of pricing below the official OPEC level to war risks. Its oil minister, Mohammad Gharazi, explained, 'We do not have discounts. The difference in the price is the extra cost resulting from the imposed condition of the war'.[41] Discounts of about $3 a barrel were offered in 1984–5 to oil buyers.[42]

Saudi Arabia, OPEC's leading oil exporter, is also thought to have exacerbated price weakness by a major barter deal agreed in July 1984: it was agreed that ten 747-300 aircraft would be delivered by the US Boeing Company in exchange for 35 million barrels. Prince Sultan Ibn Abdel-

Aziz, the defence and aviation minister, bypassed Ahmad Zaki Yamani, the minister for petroleum and mineral resources. The deal, worth $1,015 million,[43] and other barter deals by various OPEC members illustrate the rivalry between OPEC *group* interest and *national* interest as perceived by government officials. The oil received by Boeing would be disposed of in the spot market.

As sovereign states, OPEC member governments do not consider themselves bound by OPEC's resolutions on price levels and production quotas. Ecuador openly flouted its OPEC quota of 185,000 barrels per day by producing, in mid-1985, 280,000 barrels per day, with a target set at 300,000 barrels per day by the end of 1985, as acknowledged in April of that year by the president of Ecuador. The OPEC Ministerial Executive Council warned Ecuador with the sole sanction at its disposal, i.e. suspension of membership.[44] In 1985 the majority of OPEC members, according to Saudi Arabia's King Fahad, violated OPEC's resolutions with respect to production ceilings. He also gave notice that Saudi Arabia would feel free to expand its sales and reduce prices if discipline in OPEC ranks was not maintained.[45] Moreover, Saudi Arabia formally repudiated the role of 'swing producer' in 1985. The role consisted in obligating Saudi Arabia to accept proportionately greater cuts in its production share than other member countries whenever the OPEC production ceiling was lowered, as was the case in 1983–84 (See Table 11). Similarly, the role of 'swing producer' would entitle Saudi Arabia, in any eventual increase in the OPEC production ceiling, to obtain a higher proportionate increase than other OPEC members. However, the opportunity for 'upward swinging' has not yet materialised (up to publication, in 1985).

Notwithstanding OPEC's agreements on production programmes and/or official price levels, in the 1980s several member countries have been anxious to expand their sales receipts. In a buyers' market this has led to substantial, widespread overt and covert overproduction (by reference to agreed quotas) and to discounts from official prices. OPEC oil ministers have candidly admitted these 'malpractices' to be widespread.[46] Apart from persuasion, OPEC's sole sanction is the threat to exclude the non-conforming member from the Organisation. This, however, is barely effective.

Indeed, OPEC does not have the means of implementing its price-stabilisation policy. An effective cartel needs mandatory powers of controls and sanctions. The closest historical example of effective control over oil producers was the state control agencies in the United States, in particular the Texas Railroad Commission (TRC), which, from 1919 until the 1950s, enforced allowable production levels on Texan oil-fields, in consonance with similar state commissions elsewhere in the United States. The initial objective of state oil regulatory agencies in the United States was to prevent waste. Another objective was that of preventing the buyers of crude oil (who happen to be large firms in a relatively concentrated 'upstream' US oil industry) from resorting to policies of favouritism or discrimination among the numerous smaller-sized oil producers. A third objective was that of stabilising oil prices under 'market demand' rationing.[47]

Outside the centrally planned economies, which generally do not tolerate competition, public authorities in market economies have legalised or condoned cartels and/or production and export controls. Their motives are varied; they include, in particular, the desire

(a) to favour the growth of national enterprises, especially in export markets, without victimising national consumers;
(b) to enable a branch of economic activity to overcome a grave recession;
(c) to countervail monopolistic practices obtaining in other economies;
(d) to adjust supplies in response to demand, thereby avoiding excessive fluctuations in prices; and
(e) to smooth the phasing out of obsolete businesses or to facilitate their restructuring in response to changes in market and technological conditions.

In the market economies of the West one can refer to US legislation which, since 1918, has authorised US-based enterprises to set up export cartels (as per the Webb–Pomerene Act) enforceable by the courts of the country. The Australian Government has the power to impose export controls on a number of commodities, to ensure for the enterprises concerned 'fair and reasonable market prices'. According to one official source:

> The Australian Government has found that there is a need for this reserve power, particularly when situations of domestic and international over-capacity exist, and when Australian producers are in the position of dealing with single or coordinated monopoly-buying techniques which distort normal markets and depress the prices received for Australia's resources.

Other objectives of Australia's export controls are: observance of international and strategic obligations, satisfaction of essential domestic requirements and environmental protection.[48] The European Community has also resorted to cartel techniques. In particular, it resorted to monitoring curbs on 350 steel producers, with a fine of $100 per tonne on every producer who exceeded the authorised limit.[49]

In the energy market, uranium producers, with support from host governments (with the exception of the United States), have since 1972 operated an international cartel comprising eighteen state-owned and private-sector companies from Australia, Canada, France, South Africa and the United Kingdom. After 1975, companies supplying the US market were barred by US anti-trust legislation from maintaining their collaboration with the cartel. The position of the uranium cartel is constrained by three factors: first, the growing number of actual and potential uranium producers; second, countervailing co-ordination by uranium buyers; and third, absence of general support from governments (in particular, the United States) which master the necessary discipline to enforce group decisions more effectively than a private cartel.[50] Market conditions appear to have generally prevailed in uranium prices.

For another example of a cartel in the energy field, one could refer to the International Electrical Association, founded in 1930; since 1975 its headquarters have been in Switzerland (Pully). Switzerland authorises the organisation of cartels subject to certain general rules. The Association comprises some fifty European and Japanese enterprises in the electrical equipment industry covering the manufacture and installation of hydro-electric power, thermal and nuclear power turbines, generators and transformers. Its aim, as indicated by Swiss Government sources, is to reduce or eliminate their mutual competition within Third World countries. Since their activities do not affect Western European countries, or Japan, this cartel is not forbidden in these countries.[51]

Outside the energy field, production controls are prominent in several market economies, including the United States. To illustrate, the US Federal Government has been holding back the production of agricultural products to sustain their price level. Thus, in fiscal year 1978, US farmers were prompted to take about 17 million acres out of crop production (wheat and feed grain). In return, the producers were eligible for commodity loans at subsidised rates as well as for deficiency, disaster and diversion payments. The latter three payments reached over $1.8 billion in fiscal year 1978 and over $1.5 billion in fiscal year 1979.[52] In 1984 the maximum allowable plantings for wheat were 70 per cent of the average acreage planted in 1982 and 1983, and 90 per cent for feed grain (corn, sorghum, oats and barley).[53]

The United Kingdom has its own authorised cartels. This is the case with a basic industry: cement. The four domestic producers have been fixing the cement price since 1934—competing only in quality and service. The cartel's price has proved to be above world prices: West Germany's price in 1982 was 40 per cent cheaper, even for delivery to the furthest part of Britain. The British cement industry, sheltered by a relatively high price, has been less cost-efficient—especially with reference to energy input.[54]

Pricing Oil Exports

What, in the absence of competitive market pricing, should be the optimal administered enforceable price level(s)? This question has proved too difficult for OPEC to answer. Different OPEC countries have different degrees of urgency for the use of oil revenues (often encapsulated in their social discount rates). They also have different sizes and costs of proven reserves. Decisions have been the product of bargaining and compromise, with shorter-term political and economic considerations weighing heavily in the balance. Thus, in the 1980s, the price level that OPEC sought to stabilise proved unrealistic, in so far as it built up surplus productive capacity.

Besides deciding on the level of the reference or marker crude oil, it is necessary to agree on the grid of crude oil and products exported by various OPEC countries in the Eastern and Western hemispheres. Agreement was often limited to one kind of crude in a given country—that is, Saudi Arabia's Arabian Light (34°API)—among scores of alternatives. Price competition can take various forms with respect to quality and geographical differentials, rebates to erstwhile concessionaires, price discounts of petroleum products, manipulation of fiscal charges, trade credits, countertrade and government-to-government transactions. OPEC statements have acknowledged that agreement has always been difficult to reach and impossible to enforce in a continuously changing balance of supply and demand and in an environment characterised by high uncertainty.

Administrative price decisions based on the judgement of the few can thus 'overshoot' or 'undershoot' the desired optimum. This was acknowledged by the Saudi minister for petroleum and mineral resources. For example, he conceded retrospectively that the OPEC price freeze of 1976–78 was at variance with longer-term market forces and contributed to the sharp price increases of 1979–80. He commented thus after the event:

> As a result of the freeze imposed by Saudi Arabia, oil prices have more than doubled since the end of 1978. If the Western countries had been realistic and had accepted regular annual increases in the price of oil, there would not have been such sharp and sudden increases in prices, with the ensuing impact on their economies. I wish we had not frozen the price of oil in the past.[55]

This statement requires critical evaluation. The more than doubling of oil prices in 1979–80 (Table 13) was in fact provoked in large part by panic-buying stockpiling, following the interruption of Iranian oil supplies which occurred in the last months of the Shah's reign. During that period, OPEC countries raised their prices—albeit not uniformly—in the wake of price increases in the spot market. Later in 1985, the Saudi oil minister regretted the 'grave mistake' made in acceding to other OPEC members' wish to double official prices in 1979. He admitted, 'We were misled by the high increase in demand. We did not distinguish between demand and consumption—about 4 million barrels per day were going to stock. Then it came back as a source of supply'.[56] In this context it is pertinent to underline the scope for human error or misjudgement in ascertaining the

Table 13 Changes in selling price of Arabian Light crude oil*

Day/month/year	Dollars per barrel	% change	Day/month/year	Dollars per barrel	% change
1/1/70	1.39	—	1/1/74	8.32	131.11
1/6/70	1.34	(3.6)	1/3/74	9.30	11.8
30/11/70	1.41	5.2	17/5/74	9.60	3.2
15/2/71	1.68	19.2	19/9/74	9.80	2.1
1/6/71	1.74	3.6	3/10/74	10.15	3.6
1/10/71	1.79	2.9	1/11/74	10.46	3.1
1/1/72	1.75	(2.2)	1/10/75	11.51	10.0
20/1/72	1.87	6.9	1/1/77	12.09†	5.0
17/7/72	1.91	2.1	1/7/77	12.70	5.1
1/1/73	2.00	4.7	1/1/79	13.34	5.0
10/1/73	2.10	5.0	1/4/79	14.55‡	9.1
1/4/73	2.19	4.3	1/6/79	18.00	23.7
1/6/73	2.29	4.6	1/11/79	24.0	33.3
1/7/73	2.33	1.8	1/1/80	26.00	8.3
6/7/73	2.48	6.4	1/4/80	28.00	7.7
1/8/73	2.55	2.8	1/11/80	32.00	6.7
4/10/73	2.70	5.9	1/10/81	34.00	6.3
16/10/73	3.65	35.2	1/2/83	30.00	(11.8)
1/11/73	3.70	1.4	1/3/83	29.00	(3.3)
1/12/73	3.60	(2.7)	1/2/85	28.00	(3.5)

Notes:
* Prices shown here are those for Arabian Light crude, regarded since February 1971 as the benchmark for other Saudi Arabian and OPEC crude grades. They are FOB prices quoted at Ras Tanura.
† Two-tier pricing came into effect, lasting until the next price change.
‡ From the spring of 1979 until October 1981 OPEC did not have a unified pricing system.
Source: Exxon Corporation, *Middle East Oil and Gas*, Exxon Background Series, New York, December 1984, p. 27; updated till 1985 with *OPEC Bulletin*, Vienna, February 1985, p. 6.

evolution of fundamental market forces and in attempting to manage oil prices.

A number of oil-men and researchers have attributed special responsibility to Saudi Arabia in allowing oil prices to rise to the levels of 1980 and 1981, despite earlier assurances that it would prevent such increases. Indeed, Saudi Arabia—notwithstanding its large oil production capacity and reserves—refused to allow its oil production to rise further, sufficiently to prevent the price increases.[57] The Saudis then appeared constrained by two factors: first, the fear of alienating their fellow OPEC members by a very sharp increase in their market share; and second, the concern that they might endanger the long-run 'maximum efficient rate' of recovery of their oil and gas reserves, should they accelerate production from existing wells without appropriate investments. Other factors which contributed to the 1979–80 price increases were low perceived levels of commercial stocks available to oil companies, and the unwillingness of industrial countries to release oil from their strategic stocks. This provoked some companies into panic buying. The situation has to be contrasted with the much milder reaction in late 1980 at the start of the Iran–Iraq war, when stocks were larger and readily available to meet supply interruptions.

By gradually accommodating the wishes of price maximisers over 1979/81, OPEC as a group has opted for shorter-term gain against longer-term profit. Higher prices have enticed enterprises to explore for, and develop, oil deposits outside the OPEC area with a view to capturing the high profits obtaining at the production stage. Higher prices have also provoked consumers to substitute less expensive or more reliable sources of energy for OPEC oil, and to conserve energy, notably by saving on energy consumption (e.g. by reducing car trips or car speed, car pooling, greater use of public transport, reducing heating levels and improving insulation of buildings) or by improving the yield of energy-using equipment. The impact of these factors was bound to reduce, as in fact it did, what was the dominant position of power in the market of OPEC producers *vis-à-vis* non-OPEC producers (Table 14).

OPEC's price regulation has neglected flexible adjustments in response to changing market conditions. Fixed official prices have thus proved unsustainable and therefore a grave source of dissension among member countries. In periods when a buyer's market has prevailed, price rivalry among OPEC members has verged on threats of price war. These threats were voiced openly in February 1983 by OPEC ministers such as Manaa

Table 14 World oil production* (barrels daily)

	World	OPEC	OPEC/world (%)
1973	58.5	31.3	53.5
1974	58.6	31.1	53.0
1975	55.7	27.5	49.4
1976	60.1	31.1	51.7
1977	62.6	31.7	50.7
1978	63.0	30.3	48.0
1979	65.8	31.5	47.8
1980	62.7	27.4	43.7
1981	59.4	23.4	39.4
1982	57.0	19.9	35.0
1983	56.7	18.5	32.6
1984	57.8	18.3	31.8
1985	56.9	16.5	29.0

* Including natural gas liquids.

Sources: For 1973–84, *BP Statistical Review of World Energy*, London, June 1984 and June 1985, p. 5; for 1985, estimates based on the first half of the year as per: *Petroleum Economist*, London, September 1985, pp. 322–3.

Saeed Al Oteiba, UAE's minister for petroleum and mineral resources, in order to defend his country's market share.[58]

'Rational' regulation of petroleum product prices, according to the chief executive of the Saudi state-owned oil company, Petromin, should be based on the 'long-term competitiveness of the particular product in question in the market from the standpoint of substitution for other products'. In 1984 Petromin adopted a weighted formula for petroleum products (comprising spot prices and contract prices) as an optimal method of rational price regulation.[59]

Norbec, an affiliate of Petromin, was reported to have openly disregarded official OPEC benchmark prices by selling at the lower spot prices in 1984 and 1985.[60] Later in 1985, the Saudi Government agreed to sell about 800,000 b/d of crude oil to major US companies in the Aramco partnership (viz. Exxon, Chevron, Texaco) at market prices calculated on a 'netback' basis (i.e. derived from product prices after deducting transport, refining and distribution costs and normal profit margins).[61]

OPEC's Effectiveness

OPEC has been steered by different member countries over different periods of time. Those who had the 'clout' to initiate changes had the physical resources, oil or non-oil, to back their leadership role. Thus, in 1970, revolutionary Libya under Muammar al-Gadaffi dared for the first time in the history of Middle East oil concessions to impose 'unilateral' increases in the prices of, and fiscal charges on, Libyan oil exports. These increases were possible because Libya was then supplying close to 25 per cent of West European oil imports and because Libya chose to impose its will, to begin with, on the most 'vulnerable' oil company—Occidental Petroleum, which depended on Libya for oil supplies outside the United States.

In the 1970s and the early 1980s, Saudi Arabia sought to moderate price increases. Endowed with the world's largest commercial oil reserves, it was ready to back up its pricing strategy with increased oil exports. Yet the Saudi oil minister admitted that he was pressurised by other OPEC members into raising prices from $28 to $32 per barrel at the Bali Conference of December 1980, conceding in the hope of a unified OPEC stand on prices. This price increase was agreed despite the Saudi minister's firm conviction that such an increase would provoke a production surplus along with a sharp reduction in consumption. This situation, he predicted, would result in the collapse of the OPEC price increase, although it was reportedly accepted in the interests of OPEC solidarity.[62] Prices were unified at the end of October 1981 when the OPEC Conference of Ministers agreed to raise the official price of 'marker' crude to $34 a barrel.

The president of a founding member of OPEC, Venezuela, saw OPEC in 1980 as a source of power in international affairs. He expressed his views thus: 'The past 20 years of experience have shown that OPEC is a powerful instrument of world influence which no one can ignore, set aside or be unaware of in this century. The organization is unquestionably a force in power relations today'.[63] This statement, made at the height of the oil prices and OPEC euphoria appeared incongruous in 1985, when oil prices were sliding, OPEC production was practically halved and member countries in conflict or disarray.

Whether or not OPEC has been a positive factor in international relations is a question open to controversy. To some observers, OPEC has been a factor of confrontation, exacerbating already tense relations on the international scene. To others, it has proved on balance a factor of moderation, by comparison with what would have been the situation had all demands of oil-exporting states been immediately acceded to. In effect, radical national policies of selected oil-exporting countries have been tempered by group deliberation. According to a student of the subject: 'In fact, to attempt the destruction, or even weakening, of OPEC as an organisation would be folly. If OPEC did not exist, it would be in the interest of the industrial world to promote its creation in some form'.[64]

The economic ills of the industrial nations have their origins closer to home. OPEC was conveniently, and frequently, used as a scapegoat. An official of a Western industrial power called for an end to blaming OPEC for the problems faced by the world economy:

> The consumers need to stop blaming the OPEC cartel for things which in reality are determined as much by market forces as the decisions of Governments. The consumers will need to appreciate that it is not unreasonable for producers to develop depletion policies which pay regard to the needs of future generations and are designed to ease the process of adjustment by the producing country as oil becomes more scarce. Above all, the consumers need to show that they are not simply looking to OPEC to produce ever more oil on a short-term basis, with no regard to the long-term, in order to meet the growing energy needs of the industrialised economies and the rest of the world, but that they are serious about tackling the demand side of the equation, by developing positive and effective policies for energy conservation and the promotion of alternatives to oil.[65]

The outbreak of the full-scale Iran–Iraq war in September 1980 did not lead to a repetition of the price tension of 1979 which was provoked by interruptions in Iranian oil exports during the Iranian revolution. The IEA acknowledged that, besides measures adopted by its own member countries, 'responsible reactions by some of the major [OPEC] producers helped to prevent a price explosion similar to the one in 1979'.[66] These OPEC producers agreed to make up for the shortages in oil supplies caused by the paralysis of oil exports from Iraq and Iran. The supplies in question (about 3.8 million barrels per day) were largely replaced by Saudi Arabia, and to a lesser extent by Kuwait, UAE and Qatar. These countries (with some 10 million nationals, compared to 50 million for Iraq and Iran), it is worth noting, had at that time much less need for additional oil export revenues: they enjoyed much higher per capita incomes and large balance of payment surpluses[67] (see also Chapter 6).

Another indication of this 'responsibility' is that the largest oil exporter, Saudi Arabia, whose Arabian Light crude is used as a reference by a number of OPEC countries, did not sufficiently raise its oil export prices in line with those of the spot market obtaining in 1979 and 1980. Oddly enough, the United States Government had then sought the *maximum* price for its federally owned oil reserves. This happened late in 1979, when US refiners bid up to $41 per barrel for some 130,000 barrels per day of crude produced at the Elk Hills Naval Petroleum reserve in California. At that time, comparable OPEC crude was selling at much lower prices, taking into account location and quality differentials. Saudi as well as European officials protested the US auction sale 'fearing it would undermine Yamani's [the Saudi oil minister] efforts to hold down OPEC price increases'.[68] The US–OPEC price difference can be attributed to a 'security premium' in favour of US oil, feverish bidding by refiners who had encountered difficulties in procuring foreign crudes, and/or expectations of future price increases caused by potential supply shortages.

Panic-buying was not solely the lot of independent refiners. Major oil companies fell into this too, instead of destocking. Shell, British Petroleum, Exxon and France's CFP were reckoned to have paid $3–$5 per barrel more than official OPEC prices late in 1980.[69]

One scholar exposed the distortions inherent in blaming OPEC countries for the ills of Western economies. He explained the pricing policies of oil exporters thus:

> The notion that a cartel conspires to raise prices in defiance of market conditions obscures the fact that in a seller's market the sellers can expect to receive a greater share of the difference between the cost and usage value of their product. Market forces have had more to do with oil prices than OPEC. . . .
>
> Instead of allowing consumers to purchase at prices reflecting cost, producers have moved prices close to those of alternative energy sources, though not to the high usage value of oil. They have achieved a redistribution of the huge margin between producer cost and user value, a margin that used to go almost exclusively toward building the standard of living of the consuming countries.[70]

In the judgement of the author quoted, 'Excess demand has turned a buyer's market into a seller's market. Only a reduction in that excess demand can relieve the pressure for price increases. The United States, with its energy-inefficient automobiles, houses, and factories, is the number-one culprit, not OPEC'.[71] With the higher energy prices in the 1970s and early 1980s, followed by higher rates of conservation and fuel efficiency as well as by a world economic recession, demand for energy levelled off. The brunt was borne by OPEC countries which sustained the virtual halving of their oil exports over the period 1979–85.

Unlike its policy of countering excessive oil price increases in 1979 and 1980, official Saudi policy enunciated during the depressed oil market of 1985 was unwilling to prevent a hypothetical price flare-up. The Saudi oil minister openly threatened:

> If prices catch fire, we would let the fire flare for some time, because we need this sort of lesson. I don't think it is healthy for consumers to operate without stocks. I don't think the Saudi floating reserve should be looked at as a security to dampen down a flare-up in prices.[72]

OPEC's Prospects

OPEC's cohesion and survival is premised on the existence of certain conditions, chief among which are:

— the continued commercial importance of petroleum with a relatively inelastic demand;
— the concentration of resources among a few important member countries with similar large, low-cost commercial reserves; these countries perceive gains from joint action—especially in the areas of price and production;
— the absence of exacting, centralised or supranational powers in the OPEC organisation incompatible with the sovereignty of important member countries;
— the absence of virulent antagonism or countervailing/retaliatory action from one (or more) dominant world powers or from a block of oil-importing countries.

The possibility of broadening the group by including a major new exporter of the commodity in question would depend on these favourable factors prevailing. The withdrawal of a member country could prove too costly, should it provoke reprisals initiated by the low-cost, large reserve members. This was eloquently put by Nigeria's minister for petroleum and energy:

> OPEC can team up to price us out of the market and frustrate all our strategies or projections for national development. The consequent untold hardships could culminate in social discontent and possible revolution. I have often likened such a situation to the withdrawal of a person from membership of a secret society. Once in, it is advisable to stay in even if not-so-happy. Because the maverick members could be eliminated 'in the interests' of the Organization. A number of leading

> OPEC countries can easily unleash such an attack since their production price per barrel is very low. We must not forget for one moment that we are OPEC's weakest link.[73]

With the growth of other suppliers of energy prompted by higher prices, with gains in energy conservation and efficiency, and with the slowing down of world economic activity, co-operation among OPEC countries in matters of price and supply have proved much more difficult to achieve. It is unrealistic for OPEC to fix world oil prices and to enforce on its sovereign member states a quota system. These states have refused to entrust the OPEC organisation with supranational powers which would enable it to police compliance and to sanction defaulters. It is more realistic for OPEC to function as a trade association, providing members with a vehicle for discussions on supply and demand factors and on evolving a code of conduct in the area of transnational oil competition.

NON-OPEC STATES

In the mid-1980s, four non-OPEC (NOPEC) oil-exporting countries have exercised significant impact in the world oil market: Mexico, the United Kingdom, Norway and the Soviet Union. A distinction should, however, be made between Mexico and the other countries. As an oil-exporting developing country with only a moderate level of diversification in its economic structure (compared with the industrial oil-exporting countries), Mexico is, like the OPEC countries, vulnerable to reductions in oil export revenue. Mexico, as the southern neighbour of the United States and with very close US economic ties, has to be wary of offending US interests. It is well known that the 'United States Trade Act of 1975 which denied trading benefits to OPEC members effectively killed off the idea which was then being openly canvassed that Mexico should joint OPEC'.[74]

Aware of its dependence on the United States, the Mexican Government has been seeking to diversify its trading partners by endeavouring to sell oil to Japan, Canada and Europe, against advanced technology goods from these countries.[75] According to one official US source:

> The Mexican people have proved to be very nationalistic about their natural resources, especially petroleum. . . . There has been domestic opposition to the government's oil export policy, particularly exports to the United States. Many fear that increased exports of oil and gas to the United States would result in a greater US influence in the Mexican political and economic system.[76]

Although dependent on the US economy, the Mexican Government is sympathetic to OPEC price stabilisation policies. As a major oil exporter, Mexico finds these policies generally consonant with its own national interest. Available evidence suggests that Mexico was even reluctant to concede price cuts. That was the case in mid-1981 with the French state-controlled Total. In late June 1981 Total suspended their contract with the state company of Mexico, PEMEX, for lifting Mexican crude, on the grounds that it had become $3 per barrel more expensive than comparable crudes. The Mexican Government considered the Total contract to be an integral part of an economic co-operation package agreed with the French Government. The Mexican Government felt that oil lifting should not be subject to suspension in the event that short-term market forces weakened spot prices. Access to Mexican crude was negotiated in 1980 during a period of supply shortages by the conservative government of France under President Giscard d'Estaing. The French Government then wanted to diversify out of an excessive dependence on Gulf states' crudes, whose availability had been adversely affected by political factors. The French managed to negotiate several billion dollars' worth of contracts for the sale of French goods, equipment and know-how (e.g. for the construction of an underground railway system for Mexico City, nuclear power reactors, aluminium smelters and auto plants). Thus, in retaliation for the suspension of oil purchases, the Mexican Government suspended the French contracts.[77]

The French socialist government under President François Mitterand, however, assumed power later in May 1981. As a firm supporter of North–South co-operation, he chided Total's shortsighted, narrow mercantile approach. The French Government used its influence on Total to resume its transactions with PEMEX, in the interest of the two countries.[78] It is worth noting that the French state owns, as of 1981, 46.74 per cent of Total's capital and controls 50.84 per cent of its voting rights.[79]

Unlike Mexico and other oil-exporting developing countries, the United Kingdom as an industrially developed country does not share as much in OPEC's conditions and concerns. Thus, the UK authorities have adamantly refused to join any scheme of oil production controls or price stabilisation. In 1983–85 the United Kingdom was the fifth-largest producer of oil in the world, and the largest exporter of light, low-sulphur crude oil, with state-owned BNOC entitled to over 55 per cent of the UK crude oil produced from the continental shelf between 1979 and 1984 (in 1984, 1.4 million barrels per day out of 2.5 million barrels total production).[80]

On 18 February 1983 BNOC's price-cutting leadership was demonstrated when, in a buyer's market, it introduced a price cut of 10 per cent (i.e. $3 per barrel, thereby reducing its light-quality oil to $30.50 per barrel); this was immediately matched by fellow traveller Norway. A day later, Nigeria, which sells comparable crude, reduced its official price from $35.50 to $30 per barrel. Soon after, various exporters followed suit. It is significant that Britain's prime minister, Margaret Thatcher, had reportedly forewarned King Fahd of Saudi Arabia, and that the level of price cut in the British North Sea oil was 'correctly predicted' earlier by the Saudi oil minister, Ahmad Zaki Yamani.[81]

The then British energy secretary, Nigel Lawson, on a tour of Arab Gulf countries in May 1983, disclosed that his government wanted to reinforce the 'informal but valuable relations' with the key Gulf members since the OPEC 14 March 1983 agreement on voluntary constraints on members' production in order to prop up prices. He commented favourably on the fact that 'the importance of OPEC is that it became the swing producer for the world and Saudi Arabia is the swing producer of OPEC'.[82] The British Government, however, officially ruled out production cutbacks to assist OPEC's price stabilisation in the 1980s.[83] Other NOPEC producers also took advantage of the OPEC pricing umbrella, with no commitment to OPEC production controls.

Like the United Kingdom, Norway, another industrial NOPEC, has more affinity with the developed industrial countries, notably those of the International Energy Agency (IEA). According to a Norwegian who was involved in his country's oil policy:

> The United States Embassy in Oslo was particularly active in lobbying in favour of full IEA membership for Norway. The solution chosen was of Norwegian associate membership in the IEA, i.e. participating in all essential activities with full access to information, but without being pre-committed to crisis management and keeping full sovereignty over matters of energy policy in general and of petroleum policy in particular.[84]

Furthermore, Norway has cultivated friendly relations with some OPEC countries. In particular, the Norwegian Ministry of Oil and Energy joined OAPEC[85] in setting up a joint co-operation committee in October 1979. The Norwegian and Arab sides are led, respectively, by the managing director of the Norwegian state-owned national oil company, Statoil, and the secretary general of OAPEC. The purpose of the committee is to promote co-operation between Norwegian enterprises and OAPEC-sponsored groups (e.g. the Arab Maritime Petroleum Transport Company, the Arab Petroleum Service Company and the Arab Petroleum Investments Corporation).[86]

Norway's aforementioned relations with OAPEC have not constrained its freedom. Thus, Statoil initiated price reductions in North Sea oil exports in October 1984 without prior consultations with OPEC or OAPEC countries. This triggered a general reduction in prices, not only in the North Sea, but world-wide. OPEC countries were forced to readjust their own price structures. They could not prevail on Norway to rescind its decision, despite the fact that the Norwegian deputy energy minister, Arild Roedland, described the Statoil price reduction as a 'total error of judgement'.[87]

OPEC members have been forced to bear the brunt of the expansion of

NOPEC's exports. Thus, OPEC's share in world oil production has fallen drastically over recent years—from 53.9 per cent in 1973 to 31.8 per cent in 1984, and its volume of production from 31.3 million barrels per day to 18.3 million per day (see Table 14).

OPEC consultation with NOPEC producers has become imperative, and since 1982 this need has been translated into frequent contacts. Furthermore, Mexico has been invited to participate in OPEC meetings with observer status and has adopted a self-imposed 1.5 million barrels per day export ceiling.[88] The Mexican Government agreed in November 1984 to reduce its oil exports by 100,000 barrels per day in support of OPEC's decision to cut down its member countries' exports by 1.5 million barrels in order to shore up prices.[89] Egypt and Malaysia were also invited to participate in OPEC meetings and to co-operate in restraining their exports.

Even the Soviet Union received an OPEC ministerial mission in mid-1983. Led by the Algerian energy minister, Belkacem Nabi, its purpose was to invite the Soviet Government to lend its support to OPEC's efforts at stabilising prices. The Soviet response was ostensibly positive.[90] Nevertheless, Soviet oil exports have primarily responded to the country's need for foreign exchange—since the oil trade is the source of some 60 per cent of Soviet foreign-exchange earnings.[91]

In 1984 the OPEC ministerial conference entrusted the Saudi oil minister with the task of visiting NOPEC states in order to solicit their co-operation. The Saudi oil minister remarked in an interview early in May 1985 that these countries' co-operation was shortlived: they had resumed 'producing at full capacity. . . . They co-operated with us—until they relaxed, until they felt sure that OPEC is doing its job, so why not take a free ride'.[92] Excess capacity and low marginal costs tempted these producers into expanding sales, with OPEC holding the price umbrella significantly above replacement costs of the sales pushers.

Faced with a downward slide in their market share and in oil prices, OPEC members have uttered threats of 'price wars' to persuade NOPEC (in particular, the United Kingdom and Norway) to reduce their levels of production below full capacity. As the Saudi oil minister made clear: 'When we enter that war we will not create a stable lower price for oil. We will just go up and down and hit whatever floor we want to hit. I think the damage will be very serious, not only for the UK, but for banks in the US'—(which are creditors to such heavily indebted oil-exporting countries as Mexico).[93] It is doubtful whether such threats can easily be carried out. Indeed, they have had little effect in deterring North Sea producers from full capacity production and from selling at competitive prices below official OPEC prices in the buyers' market of 1981–85. Moreover, a price war such as this would hurt NOPEC as well as OPEC member countries; that is, the higher-cost producers and the heavily indebted ones (e.g. Nigeria).

Membership of OPEC has not increased since 1975. Applications were submitted by potential candidates, but founding members have exercised their right of veto to deny membership for various reasons. Thus, in 1979, the Angolan minister for petroleum openly declared 'We applied for membership as soon as our country became independent and we are looking forward to a positive decision from this Organisation'.[94] The refusal to admit Angola to OPEC is not due to its relatively modest level of exports. Two other OPEC members—Gabon and Ecuador—have a similar export level. One plausible reason is that a few major OPEC members do not want to increase the number of small exporters within their organisation and consequently to further complicate the process of attaining consensus. Another reason is the concern of a few OPEC countries about the 'radicalism' of the Angolan regime.

Outside Angola, the other developing non-OPEC countries with similarly moderate petroleum exports, as of the mid-1980s, were Bahrain, Bolivia, Brunei, Burma, Congo, Egypt, Malaysia, Oman, Peru, Syria, Trinidad and Tunisia. At this time none of these countries appeared interested in joining OPEC since this would mean complying with OPEC production and price control schemes. NOPEC countries have generally benefited from policies espoused by OPEC countries which in the 1980s accepted reductions in their production to sustain world oil prices. Nevertheless, OPEC's capacity to assume price leadership in that period, by comparison with the 1970s, proved elusive.

NOTES

1. See Francisco R. Parra, 'Venezuela, Oil and the Middle East', *Arab Gulf Journal*, London, October 1983, pp. 49–56.
2. For the genesis of OPEC, details in Z. Mikdashi, *The Community of Oil Exporting Countries*, Allen and Unwin Ltd., London and Cornell University, Ithaca, New York, 1972.
3. Ali A. Attiga, 'Economic Development of Oil-Producing Countries', *OPEC Bulletin*, Vienna, November 1981, p. 10.
4. See also 'Algerian Press Criticizes Saudi Role in OPEC', *Middle East Economic Survey* (*MEES*), 18 March 1985, pp. A5–A7.
5. James E. Akins, 'US Energy and Middle East Oil and Politics', in *Natural Resource Policies in Economic Development and International Cooperation*, **1**, Madison, University of Wisconsin, 1978, p. 356.
6. Farouk M. Akhdar, Secretary-General of Saudi Arabia's Royal Commission for Jubail and Yanbu, 'The Philosophy of Saudi Arabia's Industrialization Policy', *MEES Supplement*, 24 November 1980, p. 1.
7. 'Nazer Says Half Present Oil Output Would Meet Financial Needs', *MEES*, 27 April 1981, p. 1.
8. 'Yamani admits to promoting oil glut to stabilise prices', *Financial Times*, London, 21 April 1981, p. 1, and *MEES Supplement*, 27 April 1981.
9. See Bluford Putman and Lee Thomas, 'Why the Saudis Must Cut the Oil Price', *Euromoney*, London, February 1985, pp. 40–1.
10. Interview in *MEES*, 20 May 1985, p. A4.

11. See, for example, *Critical Factors Affecting Saudi Arabia's Decisions*, Report to the Congress of the United States by the Controller General, Washington DC, 12 May 1978, in particular pp. 56–7.
12. See 'Reagan says US would bar a takeover in Saudi Arabia that imperiled flow of oil', *New York Times*, 2 October 1981, p. 1.
13. Ahmad Zaki Yamani, at the Foreign Policy Association in New York on 23 April 1981, reproduced by *MEES*, 11 May 1981, p. ii.
14. US General Accounting Office, *US Role as Contracting Agent for the US–Saudi Arabian Joint Commission on Economic Co-operation*, Washington DC, 14 May 1984, p. 2.
15. US General Accounting Office, *Reimbursement of Federal Employees' Salaries and Benefits by Saudi Arabia*, Washington DC, 21 October 1982, pp. i and ii.
16. James E. Akins, 'The Influence of Politics on Oil Pricing and Production Policies', in symposium on 'Oil and Money in the Eighties', held in London on 28–29 September 1981 by *International Herald Tribune* (*IHT*) and *The Oil Daily*, reproduced in *Arab Oil and Gas*, 16 October 1981, p. 26.
17. See Michael Petrie-Ritchie, 'Saudi Offset Programme Takes Off', *Middle East Economic Digest* (*MEED*), London, 27 July 1984, p. 25; 'Riyadh seeks offset spending on AWACS', *Financial Times*, London, 2 February 1984, p. 1; and 'Offset Investments in Saudi Arabia to Promote Technology Transfer', *MEES*, 24 May 1984, pp. B1–B2.
18. See, for example, Sultan J. Jaber, 'American Security Measures in the Arab Gulf', *Arab Gulf Journal*, London, October 1985, pp. 31–8.
19. See 'Le GIGN: une force d'intervention en cas de prises d'otages', *Le Monde*, Paris, 11–12 October 1981, p. 7.
20. US General Accounting Office, *US Role as Contracting Agent for the US–Saudi Arabian Joint Commission on Economic Co-operation*, op. cit.
21. Islam's two major sects are the Sunni and the Shiah. Most Muslim nations' populations belong predominantly to the former, while the majority of Iranians and Iraqis belong to the latter.
22. Godfrey Jansen, 'The Gulf War: The Contest Continues', *Third World Quarterly*, London, October 1984, p. 951; also, 'Gulf War: you started it, not me', *The Economist*, London, 6 April 1985, pp. 47–8.
23. See also Vahe Petrossian, 'The Gulf War', *The World Today*, Royal Institute of International Affairs, London, November 1980, pp. 415–17.
24. *MEES*, 5 November 1984, p. A4 and 15 July 1985 (quotations from Iraqi oil official), p. A7.
25. *Arab Oil and Gas*, Paris, 1 February 1983, p. 4.
26. OAPEC's membership included in 1985: Algeria, Bahrain, Egypt (whose membership was suspended upon Egypt's unilateral signature of a peace treaty with Israel in 1979), Iraq, Kuwait, Libya, Qatar, Saudi Arabia, Syria, Tunisia, United Arab Emirates.
27. 'Syrian Oil Minister Confirms that Closure of Iraqi Pipelines was Political Act', *MEES*, 21 June 1982, pp. 3–4; and 'Closure of Iraqi pipe is political not contractual dispute—Syria', *Saudi Business*, Jeddah, 20 August 1983, p. 9.
28. See Patrick Cockburn, 'Iran to Give Syria 1 m. tons of oil', *Financial Times*, 7 April 1983, p. 5.
29. For a genesis of GCC, see *Digest—The Gulf Co-operation Council*, Kuwait News Agency, May 1981.
30. Hazem El-Beblawi, *The Oil Decade: An Appraisal in Perspective*, the Industrial Bank of Kuwait, September 1983, p. 33.
31. Roger Matthews, 'The long war with Iran–Iraq: now the debts pile up', *Financial Times*, London, 26 March 1982, p. 17.
32. See, for example, Ian Seymour, *OPEC—Instrument of Change* (published for OPEC), London, Macmillan, 1980.
33. Stocking, op. cit., p. 361.

34. Oystein Noreng, 'The Western World and the Oil Exporters: An Historical Analogy', *OPEC Review*, Vienna, Autumn 1979, p. 47.
35. Marginal cost in the production of an exhaustible resource comprises (i) extraction costs and (ii) user costs (i.e. foregone profits associated with selling a unit of the exhaustible resource in the current period, instead of at a future period). See David J. Teece, 'Assessing OPEC's Pricing Policies', *California Management Review*, University of California, Autumn 1983, p. 71.
36. Ali Khalifa al-Sabah, Kuwait's oil minister, in *MEES*, 8 June 1981, p. vi.
37. Food and Agriculture Organisation (FAO) of the United Nations, *Commodity Review and Outlook: 1980–81*, Rome, 1981, p. 100.
38. 'Oil industry warned over price cutting', *Financial Times*, London, 27 January 1983, p. 18.
39. 'Saudi Arabia Lent $400 million to Nigeria in 1982', *Arab Oil and Gas*, 1 August 1983, p. 24.
40. Richard Johns, 'The oil market—a chance for stability, but the doubts remain', *Financial Times*, 6 April 1983, p. 15.
41. *MEES*, 27 December 1982, p. 2.
42. 'A Japanese casualty in the Gulf War', *The Economist*, 14 July 1984, p. 68; and 'Gulf War—bullseye', *The Economist*, 28 September 1985, p. 51.
43. See Michael Petrie-Ritchie, 'Oil, Aircraft and Emergencies: Saudi's Deal with Boeing', *MEED*, 27 July 1984, p. 26.
44. *MEES*, 20 May 1985, p. A5.
45. See *MEES*, 10 June 1985, p. A2; and 'Fahad Defends Kingdom's Oil Policy', *Saudi Gazette*, Jeddah, 24 September 1985, p. 1.
46. For example, Tam David-West of Nigeria in *MEES*, 15 July 1985, pp. D2–D5.
47. See Stephen L. MacDonald, *Petroleum Conservation in the United States: An Economic Analysis*, Baltimore, Johns Hopkins Press, 1971.
48. Australian Department of Trade and Resources, *Australia's Mineral Resources, Development and Policies*, Canberra, 1980, p. 13.
49. 'How the tonnage monitors operate', *Financial Times*, London, 29 January 1981, p. 2.
50. See Marian Radetzki, 'International Uranium: Economic and Political Instability in a Strategic Commodity Market', *CIPEC Quarterly Review*, Paris, October–December 1980, p. 1; and 'Cartel de l'uranium: six sociétés canadiennes sont inculpées d'entente illégale', *Le Monde*, Paris, 12–13 July 1981, p. 15.
51. See, for example, 'Un puissant cartel international qui a son siège à Pully', *Tribune Le Matin*, Lausanne, 5 February 1983, p. 3.
52. US General Accounting Office, *Agriculture's Set-Aside Programs Should Be Improved*, The Comptroller General Report to the Congress, Washington DC, 11 January 1980, pp. i and 2.
53. Gary L. Benjamin, '1984 Government Farm Programs', *Agricultural Letter*, Federal Reserve Bank of Chicago, 3 February 1984, pp. 2 and 3.
54. See 'Britain's cement market—a concrete case against the cartel', *The Economist*, 5 February 1983, pp. 70–1.
55. Ahmad Zaki Yamani, 'The Impact of Oil in International Politics' in *MEES Supplement*, 5 May 1980, p. 6.
56. Ahmad Zaki Yamani in *MEES*, 7 January 1985, p. D8.
57. See, for example, M. A. Adelman, 'Coping with Supply Insecurity', *Energy Journal*, Cambridge, Mass., April 1982, pp. 6 and 10.
58. Véronique Maurus, 'La crise pétrolière', *Le Monde*, 1 March 1983, p. 43.
59. Abdulhady Taher, interview with *MEES*, 29 October 1984, pp. A4–A5.
60. See, for example, *Arab Oil and Gas*, Paris, 1 March 1985, p. 3.
61. See 'Saudi Production Rises to 3m b/d in Late September', *MEES*, 30 September 1985, p. A8.

62. Interview with Yamani, reproduced by *MEES*, 14 February 1983, pp. A2–A4.
63. Speech by H. E. Luis Herrera Campins at OPEC Secretariat, in *OPEC Bulletin Supplement*, Vienna, 24–31 March 1980, p. ii.
64. Edith Penrose, 'OPEC's Importance in the World Oil Industry', *International Affairs*, Royal Institute of International Affairs, London, January 1979, p. 18.
65. David Howell, Secretary of State for Energy, United Kingdom, 'Towards a New Interdependence' at OPEC Seminar on *OPEC and Future Energy Markets*, Vienna, October 1979, p. 3.
66. International Energy Agency, *Energy Policies and Programmes of IEA Countries, 1984 Review*, OECD, Paris 1985, p. 17.
67. 'Meetings in Progress to Arrange Allocation in Incremental Gulf Crude Supplies', *MEES*, 27 October 1980, pp. 1–2.
68. 'A Sure Oil Supply—At a Price', *Business Week*, New York, 25 February 1980.
69. See, for example, 'The price of war', *The Economist*, 15 November 1980.
70. Bruce R. Scott, Professor of Business Administration at the Harvard Business School, 'OPEC, The American Scapegoat', *Harvard Business Review*, Boston, January–February 1981, p. 14.
71. Ibid., p. 28.
72. Ahmed Zaki Yamani, 'Serving OPEC Member Countries' Best Interests', *OPEC Bulletin*, Vienna, July–August 1985, p. 16.
73. Tam David-West, 'Oil Politics and Development in Nigeria', *OPEC Bulletin*, Vienna, May 1985, p. 16.
74. George Philip, 'Mexican Oil and Gas: The Politics of a New Resource', *International Affairs*, Royal Institute of International Affairs, London, Summer 1980, pp. 480–1.
75. Ibid., p. 481.
76. Controller General of the United States, *Prospects for a Stronger United States–Mexico Energy Relationship*, Report to the Congress, Washington DC, 1 May 1980, p. 34.
77. See 'La suspension du contrat mexicain de la CFP: une faute grave de la France', *Bulletin de l'industrie pétrolière*, Paris, 7 July 1981.
78. 'La "crise" entre Paris et Mexico—Un incident de paracours salutaire', *Le Monde*, Paris, 12–13 July 1981, p. 13.
79. *Bulletin de l'industrie pétrolière*, Paris, 21 December 1981, p. 2.
80. W. I. M. Gorskirk, Chief Executive in BNOC, *Report and Account for 1984*, London, 29 May 1985, pp. 5–9.
81. See Véronique Maurus, 'La baisse du prix du pétrole', *Le Monde*, Paris, 20–1 February 1983, pp. 1 and 7.
82. 'British Energy Secretary Underlines OPEC's "Stabilising" Role', *Arab Oil and Gas*, Paris, 16 May 1983, pp. 6–7.
83. See, for example, Britain's Minister of State for Energy (Alick Buchanan-Smith), statement to Parliament on 16 July 1985, cited by *MEES*, 22 July 1985, p. A4.
84. Oystein Noreng, 'Friends or Fellow Travellers', *OPEC Review*, Vienna, Autumn 1979, p. 327.
85. In addition to Arab OPEC states, OAPEC comprises two other minor Arab exporters: Bahrain and Syria (Egypt's membership was suspended in 1979).
86. *OAPEC Bulletin*, Kuwait, November 1979, p. 1 (in Arabic).
87. 'Political Row in Norway', *MEES*, 19 November 1984, pp. A10–11.
88. 'To reinforce its position in the market, OPEC looks for closer ties with other exporters such as Mexico, Oman and Egypt', *Arab Oil and Gas*, Paris, 16 July 1984, p. 4.
89. Joint statement by Mexican Energy Minister Francisco Labastida and PEMEX's Director-General Mario Ramon Beteta issued on 1 November and reproduced in *MEES*, 5 November 1984, pp. A6–A8.
90. Interview with Belkacem Nabi in *Le Monde*, Paris, 10 June 1983, p. 28.

91. See Isabel Gorst, 'Soviet Union, Striving to Maintain Oil Exports', *Petroleum Economics*, London, February 1984, pp. 61 and 62.
92. Ahmad Zaki Yamani in *MEES*, 6 May 1985, p. A7.
93. Ahmad Zaki Yamani in *MEES*, 1 January 1985, p. D6.
94. Jorge de Morais, address presented to the OPEC Seminar on 'OPEC and Future Energy Markets', Vienna, October 1979, p. 1.

4 Industrial States' Energy Interests and Policies—An Assessment

OBJECTIVES

The energy policies of energy-deficient countries have generally been guided, albeit with varying degrees of concern, by diverse sets of objectives. Chief among these objectives are:

— enhancement of energy supplies' 'security' (usually ascertained by reference to criteria of reliability and sufficiency of energy flows);
— reduction of price volatility (notably by attaining a stable price level which reflects the normal scarcity, in contrast to an artificially constrained or 'monopoly' value of a given energy supply);
— stimulation of economic activity; avoidance of waste (through conservation and improved efficiency of energy-using equipment);
— the protection of the environment;
— minimisation of energy costs (e.g. for reasons of equity consideration for the benefit of the poorer sections of society).

National interests in the energy field are not always consonant or readily reconcilable. To illustrate, a policy of cost reduction in energy development (e.g. through open-cast coal-mining) could well conflict with environmental protection (e.g. preservation of the landscape). Also, the reduction in energy costs to the national economy (e.g. by importing oil from low-cost but unstable areas) could be at variance with the development of higher-cost alternative sources of energy (notably domestic sources) as a means of diversifying supplies and enhancing national security.

Expediency or sectional interest could also be at variance with longer-term price stability. To illustrate this, one might refer to a reported Carter Administration concern about winning the support of electors living in the colder parts of the United States, through a \$5 per barrel entitlement on heating oil imports, granted up to 31 October 1979. This is equivalent to a subsidy to increase oil imports into the United States,[1] notwithstanding the fact that US officials had criticised West Germany and Italy for buying oil in excess of agreed targets during that period.[2] It is important to note that official policies are often the outcome of compromises amongst competing internal and external interests and pressures. While it is difficult to assess the impact of policies on the efficient/safe development and use of energy resources over the longer term, a policy of artificially holding energy prices either up or down by legislation would therefore be costly to the economy: it would result in a misallocation of resources, to an excessive stimulation/depression of energy demand and would eventually lead to a price 'shock'. One can illustrate this by reference to the 1973–81 regulation on domestic oil prices in the United States, which held well-head prices on 'old' oil (produced from wells in operation prior to August 1973) below world market prices. This regulation has proved costly to the US economy. One author marshalled evidence to show that it 'discouraged production, failed to retard consumption, increased the level of imports, magnified the international payments deficit and contributed to the decline in the exchange rate of the dollar'.[3]

Energy-deficient countries have primarily been worried about what they perceived to be excessive dependence on foreign supplies. Industrial countries have sought to alleviate their vulnerability *vis-à-vis* energy imports, especially with respect to the costs that disruptions could impose on the national economy. These costs can show up directly or indirectly, notably through loss of production and employment, long lead-times needed by consumers and industry, to switch to equipment and machines using more reliable fuel substitutes, higher prices and socio-political problems.[4] Some analysts have attempted to estimate the price and other economic effects of oil supply disruptions. The latter depend on a number of key variables, chief among which are the size of the disruption and its duration, the size of excess capacity readily available internationally for production and export, the size of inventories and the willingness of companies and governments to draw on them to counter 'panicky' purchases. Depending on the state of these factors, the price impact of disruption in supply can be:

— negligible: when the disruption is relatively moderate and/or surplus supply capacity is readily susceptible to prompt response;
— considerable: if the disruption is large and/or if surplus capacity is unavailable.

To fend off costly import disruptions, a number of strategies have been resorted to:

— the accumulation of inventories of energy products, including oil, for emergency periods;
— the reduction of energy imports;
— the expansion of the development of domestic energy supplies;
— the increase in energy conservation programmes;
— improvement in the energy efficiency of equipment and machines;
— the diversification of foreign sources of supply.

An evaluation of major approaches adopted by industrial countries to meet the energy challenge is attempted in the following section.

DEMAND–SUPPLY MANAGEMENT

Public authorities have been at considerable pains to influence the patterns of energy use and production. In the 1970s, the demand for oil products in the short term proved relatively inelastic to price rises and certainly to moral persuasion (as in conservation publicity campaigns). This is largely due to the extensive use of equipment fuelled by petroleum products—equipment which cannot be readily replaced.

Higher taxes, administrative restrictions or interdictions on the consumption of certain energy products and appropriate incentives or subsidies in favour of conservation or substitutes have been used. With higher prices, oil consumption per unit of GNP is likely to be significantly reduced after allowing for time in the development of oil-efficient equipment and techniques.

Diversification and Incentives

Large price increases in natural crude in the 1970s and early 1980s have led to substantial investment in research into alternative sources of energy, as well as additional expenditure on oil exploration outside the OPEC area. The period of gestation of such investments normally extends over several years, to judge from past and current experience. Imports from 'safe' producers may carry an extra cost which represents for buyers 'insurance' and/or 'scarcity' premiums aimed at reducing dependence on oil regions bedevilled by crises in a world market characterised by disruptions.

Various press reports, based on trade statistics, estimate that this extra cost amounted to about $2 per barrel in July 1979 at the height of the anti-Shah revolution in Iran, when oil exports from that country fell considerably.[5] In fact the $2-per-barrel premium was primarily attributable to higher spot prices paid on arm's-length sales in the absence of oil available at official prices (see Table 7).

Governments in various countries have supported the development of both renewable and non-renewable domestic sources of energy. Thus, subsidies to national coal production have long been made by the European countries and Japan. Direct subsidies in EEC countries for the production of coal in 1983 were highest in France (£19.19 per tonne), followed by Belgium (£15.47), West Germany (£12.06) and the UK (£9.41 for 1983–84). These government subsidies have generally been rising, either in response to less accessible/lower quality coal deposits, and/or to higher workers' pay claims. Using March 1984 price levels, the UK National Coal Board total losses (before accounting for 'deficient grants') rose from £546 million in 1981 to £716 in 1982, £818 in 1983, and £1,198 million in 1984. The leap in 1984 costs compared with those of 1983 was largely due to an industrial dispute; the latter developed into a protracted strike after March 1984 and contributed to a sharp fall in UK hard-coal production (from 122 million tonnes in 1983 to 51 million in 1984).[6] The nationalised Charbonnages de France encountered similar problems, with costs, losses and state aid rising considerably between 1979 and 1983 (Table 15).

United States' authorities have agreed to assist in the high-risk research-and-development programmes for new energy supplies. This has arisen through a number of measures covering fiscal advantages, subsidies, government guarantees, favourable credit facilities, or technical assistance to private-sector enterprises. Thus the United States Synthetic Fuel Corporation (SFC) was created in 1980 to underwrite up to 75 per cent of the eligible project costs (through purchase agreements, price guarantees and loans). In accordance with the Energy Security Act (Public Law No. 96–294, 30 June 1980), SFC could ultimately dispose of up to $88 billion in alternative fuel development. Its target was originally set at 500,000 barrels of crude-oil equivalent per day by 1987, and 2 million barrels per day by 1992.[7]

Table 15 Data on French coal production,. 1979–1983

	Extraction cost (francs/tonne)	Loss (francs/tonne)	Total state aid (millions of francs)
1979	367	113	3,985
1980	417	102	4,001
1981	467	51	4,097
1982	592	166	5,902
1983	631	194	6,435

Source: Charbonnages de France, *Rapport d'activité au 31 decembre 1983*, Paris, p. 9.

With a reduced budget of $14.8 billion for the year 1984/85, the Corporation's scope covered the support of projects which could produce fuels in lieu of natural crude oil—from shale, tar sands, coal, peat, heavy oil and other raw materials. The subsidy factor per unit of synthetic fuel was considerable; it reached up to twice the current price of natural crude oil. By the mid-1980s the US Government favoured limiting the authority and scope of operations of the SFC.[8]

Nuclear power had been perceived as a major opportunity for energy deficient countries to diversify outside petroleum, thereby enhancing the security of their energy supplies. Nuclear fuel happens to represent a small element in the cost of electricity, as compared with other sources of power generation. As such, nuclear electricity is less vulnerable to the vicissitudes of supply interruptions and price fluctuations. However, producers of nuclear fuel and technology (including the treatment of nuclear waste) happen to be located in just a few industrial countries. The governments of these countries have sought control over the immediate, intermediate and ultimate uses of their nuclear fuels, equipment and processes. Their concern is to avert the risk of dissemination of nuclear fuel and technology for hostile or military purposes. The control is thought too exacting by a neutral Western country (i.e. Switzerland) which deems these controls to be a violation of national sovereignty.[9] Switzerland was even 'punished' by the imposition of an American embargo on nuclear fuel, reportedly due to its sale of nuclear technology to Pakistan.[10] West Europeans had sought to develop their own process of uranium enrichment to avoid dependence on US supplies.[11]

The policy of energy diversification has extended to tapping energy from agricultural produce and waste. One approach has been to blend gasoline from petroleum with alcohol derived from farm products (known as ethanol), often in a ratio of 9 to 1, which gives a 'gasohol' mix. In 1982 the US Federal Government provided 4-cents-a-gallon subsidy to gasohol. In addition, some twenty-one states exempt gasohol from the full rate of state tax levied on gasoline—ranging from 1 cent to 9 cents per gallon. If the average is taken as 5 cents per gallon, since each gallon of gasohol is one-tenth ethanol, the state-tax exemption amounts to a 50-cents-a-gallon subsidy on ethanol, and the federal subsidy to 40-cents-a-gallon. As one ethanol producer commented, 'the dependence on subsidies clouds any future decision' regarding investments in grain-alcohol plants. For reference, the cost of a gallon of unleaded gasoline at the refinery amounted to about $1 in 1982.[12] The policy of 'fuel from agriculture' has so far (1985) proved insignificant in terms of its impact on the US energy market as a whole.

Excise taxes on energy products or services are bound to dampen global energy consumption, encourage conservation and promote greater energy efficiency. In so far as some energy products and their uses are taxed more than others, the mix of energy products is bound to favour those with lesser tax burdens—assuming comparable cost levels.

Conservation

Various measures have been adopted to promote the rationalisation of energy use and to encourage users to reduce waste. Among these measures are those belonging to the 'voluntary' approach. This consists in persuading users (businesses or consumers) mostly by publicity campaigns or by practical advice and guidance. More effective levers are found in the mandatory approach. These cover legislation (e.g. speed limits), tax incentives for energy savers (e.g. insulation of buildings) and/or penalties for violaters. A third approach is for goverment to underwrite the development of energy conservation technology.

The oil price increases of 1973–74 and 1979–80 have provoked strong movement towards conservation and the restructuring of industrial activity, as well as a renewed search for non-oil energy sources. New technologies and production processes have been introduced in response to the post-1973, high-cost energy era. Consequently, the OECD energy-intensity index (primary energy demand divided by an index of real GDP) fell by an estimated 18 per cent between 1970 and 1982, while the OECD oil-intensity index fell by a much greater factor of 27 per cent.[13]

Japan is a well-known leader in energy conservation and efficiency among the industrial countries. Improved energy yields have covered the production of basic goods (such as steel and cement) and services from various equipment (e.g. in home appliances and transport vehicles). Table 16 illustrates the importance of conservation between 1973 and 1982, during the upward swing in energy prices. Energy use per unit product fell considerably; the fall was mostly borne by oil, with other sources of energy improving their relative position (Table 16).

The United States' economy, the world's largest user of energy, also achieved a considerable fall in energy consumption per unit of GNP between 1975 and 1985 (Figure 3). This trend, discounting seasonal fluctuations, can be largely attributed to increases in the price of energy. These price increases have dampened energy consumption per constant dollar of GNP for a number of reasons—mainly conservation, improved energy efficiency and structural changes (i.e. a movement away from energy-intensive industries).

Strategic Reserves

The adverse effects of interruptions in supply can be reduced by the use of stand-by reserves in oil-consuming countries—whether these arise from the availability of readily producible, discovered proven resources (oil, gas, coal, etc.) or from the accumulation of stocks in man-made tanks or man-filled underground reservoirs. These reserves, whenever capable of sustaining normal consumption over several months (say six), would blunt the impact of interruptions and deter the use, or the threat of use, of embargoes.

A few Western countries endowed with hydrocarbons, in periods of apparent long-term energy shortage, have sought to decelerate the development and production of their oil or gas fields, which are seen as strategic reserves for emergency periods. This is true of Norway and the Netherlands.[14] Also, the British Government considered delaying the depletion of domestic oil reserves in 1980 in the interests of preserving energy resources for the national economy. This policy was, however, considered to be at variance with that of the UK's energy-importing partners in the EEC or in the OECD—countries which were anxious to ease pressures on world oil supplies.[15]

The policy of accumulating strategic oil reserves has mostly been

Table 16 Consumption per production unit in major Japanese industries (FY 1982/FY 1973)

Industry	Improvement in energy consumption per unit product	Improvement in oil consumption per unit product
Steel	85.0	32.0
Plate glass	70.7	65.0
Cement	71.9	3.7
Pulp & paper	91.1	64.1
Textile dyeing	73.1	64.7

Note: FY 1973 = 100.
Source: Ministry of International Trade and Industry (MITI)—Agency of Natural Resources and Energy, *Energy in Japan: Facts and Figures*, Tokyo, April 1984, p. 18.

adopted by the industrial countries. The US Strategic Petroleum Reserve contained nearly 490 million barrels at the end of September 1985 (equivalent to over a hundred days of US crude and petroleum products net imports).[16] The United States has long espoused the policy of strategic stockpiling of commodities. The massive accumulations by the United States in their strategic stockpiling of commodities in the early 1950s during the Korean war were subject to wide criticism. According to one informed observer, strategic stockpiling:

> has been used for different purposes—sometimes to depress prices or as a blackmail weapon. . . . [It] has cost the American taxpayer a great deal of

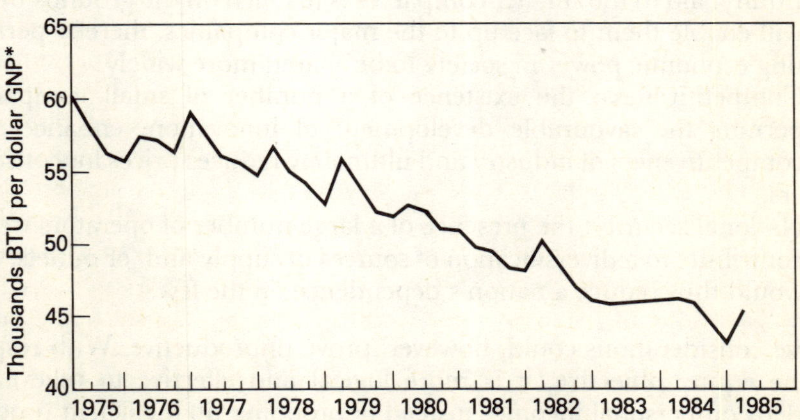

Figure 3. United States energy–GNP ratio.
Source: US Department of Energy, *Monthly Energy Review*, reproduced in *Petroleum Monthly*, August 1985, New York, Salomon Brothers Inc., p. 26.

money and has been a general disruptive influence on world markets. The indiscriminate method of panic-buying in the 1950s created chaos and resulted in the US having to pay inflated prices for materials which have then been stored at great cost for many years.[17]

As another official US study has shown, panic-stockpiling during the Iranian oil disruption of 1979 exacerbated the shortage and provoked price increases—at a time when existing excess capacity was sufficient to replace most of the lost production.[18] IEA countries have addressed this problem by agreeing to quieten the market in periods of supply interruption. A number of measures could be envisaged:

— encouraging oil companies to draw down on their stocks;
— discouraging 'undesirable' purchases (e.g. certain foreign parties or speculative trading), and holding down demand;
— encouraging greater indigenous energy production within the IEA area;
— reducing supply disparities among countries and/or companies (large and small) in order to avert market tensions.[19]

Market Structure and Regulations

Government policies in market economies have influenced industrial organisation, often in response to sectional interests. This is the case where governments have protected small companies *vis-à-vis* market domination by major companies. The rationale for supporting smaller companies is based mainly on three considerations:

— Equity: aid to the smaller companies is justified on the grounds that it will enable them to face up to the major companies, thereby permitting economic power in society to be shared more widely.
— Competitiveness: the existence of a number of small companies permits the favourable development of innovation, enhances the competitiveness of industry and ultimately reduces prices for consumers.
— National security: the presence of a large number of operators would contribute to a diversification of sources of supply and/or outlets, and would thus reduce a nation's dependence on the few.

These considerations could, however, prove unproductive. With respect to the equity objective, it is more logical and effective to relieve tax burdens on personal incomes instead of propping up inefficient production units. As regards competitiveness, most aspects of the oil industry are capital intensive. Accordingly, economies of scale should be allowed to attain their optimal level in the interest of consumers, instead of forcing a split in the market into smaller, high-cost units. Moreover, national security could well be served by the existence of a few large enterprises which have sufficiently diversified international interests, economic weight and bargaining power to obtain better terms for the national economy than smaller units.

A government bias towards small units in the oil industry can be illustrated by reference to the United States where, over 1973–81, small refining firms obtained mandatory allocations of crude-oil supplies or cash to reduce their crude-oil acquisition costs to a level enjoyed by the major refining companies or the integrated companies with access to lower-cost crude oil supplies. The US Government thus enforced a system of 'entitlements' to enable small refining firms to obtain, in this period of relative crude-oil shortages and soaring spot prices, supplies at (or cash compensation by reference to) the average price that the larger refineries with access to crude oil had paid for their supplies. Over this period the larger refineries had access to domestic crude, priced by law (Emergency Petroleum Allocation Act, 1973) below world market prices (excluding stripper wells, defined as those producing less than ten barrels a day). Price ceilings on domestically produced crude oil were considered an appropriate way of attaining a number of objectives:

- controlling the general inflationary impact of the world oil price increases of the 1970s;
- protecting US consumers—in particular those with relatively low purchasing power—from the rising cost of oil products;
- preventing the US domestic oil industry from deriving additional profits resulting from price increases provoked by external factors.[20]

Moreover, small refineries received 'a greater than proportional share of entitlements according to a sliding scale, based on the total capacity of the company owning the refinery. Large refiners who receive a lower share of entitlements pay for this small refiner bias'.[21] In 1980 the subsidy was estimated to range from $13 million to $31 million for small refiners in the category of 10,000 to 50,000 barrels per day, equal to over half the initial investment of such refineries.[22] With the phasing out of entitlements in the early 1980s, marginal refineries were forced to close down, with their losses further increased by a highly competitive market and shrinking demand.

Commenting on the effectiveness of US energy regulation, available evidence suggests that regulations were introduced in response to multiple interests and aims. In imposing regulations, US policy measures have proved inconsistent, with confusing and disappointing results. In general, government policies turn out to be politically inspired compromises in response to prevailing conditions and events.[23]

The fundamental question remains: what mix of government intervention and a free market is likely to produce greater public benefits, and in what areas? The answer will primarily depend on the political system in each country and the socio-economic context.

Thus, in the United States, administrations formed by the Democratic Party, including that of President Carter (1976–79), have been anxious to protect smaller operators. By comparison, the Republican Administration of President Reagan (1979–85) favours fewer government regulations, even if it leads to a higher degree of industrial concentration. The wave of mergers in the United States in the first half of the 1980s should be seen in the context of the Government's policy of deregulation. The subject of mergers is analysed in Chapter 5.

The Japanese Government is thought to have generally favoured the 'mitigation of the excessive competition'. Its tools for achieving that objective have largely been the tariff quota system, administative prescriptions and financial/fiscal measures introduced by the Ministry of International Trade and Industry (MITI) and the Ministry of Finance. MITI's influence has been most effective in restructuring Japanese industry in the 1980s. It obtained a 'disposal of excess capacity' and the integration of refiners'/marketers' activities (see also the second section of Chapter 1).[24] This consolidation should further improve the bargaining power of the country concerned *vis-à-vis* the outside world.

Compared to smaller nations, major industrial countries can manage, thanks to their superior bargaining power, to extract attractive terms. This was the case, *inter alia*, of the United States Government which, in August 1982, obtained Mexican oil for its Strategic Petroleum Reserve at 'an advantageous price . . . below market'.[25] The purchase was part of an arrangement to help finance Mexico after it had declared its inability to honour its external financial obligations.

Regarding the behaviour of oil companies when confronting government regulations, we may cite the candid comments of the chief executive of the world's largest private company. He admitted, 'the industry, for its part, has not always recognised that government has a legitimate role in energy decision-making. It [the industry] has tended to be unduly rigid on politically tough legislation and regulatory issues, perhaps through insensitivity to the special problems faced by the government'.[26]

GOVERNMENT AGENCIES

Several governments have not been content with influencing the international petroleum industry by means of regulation, constraints or incentives. They have moved into direct involvement by encouraging the establishment and growth of national state-owned enterprises. This strategy is based on the assumption that an operational role for government in the petroleum industry is more effective in fulfilling national objectives and thereby protecting national interests. Moreover, in a market dominated by foreign companies whose terms are deemed unacceptable by host governments, and in the absence of private domestic capital and interest, governments in market economies have launched their own state-owned or state-supported oil companies.

Among the first to seek such government involvement, the British Government bought a controlling interest in the first major oil-producing concession in the Middle East in 1914—that of William Knox D'Arcy in Iran, operating under the name of the Anglo-Persian Oil Co. Ltd. (This name was changed on 28 June 1935 to the Anglo-Iranian Oil Co. Ltd., and on 17 December 1954 to the British Petroleum Co.)

Nearly all governments of private market economies or mixed economies have at one time or another espoused the strategy of state-owned or state-supported companies to further national objectives. Notable exceptions are those of the United States and Australia.

One can illustrate government objectives in creating a state agency by reference to an industrial market economy with large energy reserves. This is true of Canada, whose Government created Petro-Canada in 1975, with the goal of achieving secure/economic energy supplies and maximum benefits from the development of national energy resources. In particular, the Petro-Canada Act of Parliament gave this national corporation the following mandate:

> to explore for hydrocarbon deposits, to negotiate for and acquire petroleum products from abroad to assure a continuity of supply for the needs of Canada, to develop and exploit deposits of hydrocarbons within and without Canada in the interest of Canada, to carry out research and development projects in relation to hydrocarbons and other fuels, and to engage in exploration for, and the production, distribution, refining and marketing of, fuels.[27]

The Canadian Government, under the Liberal leadership of Pierre Trudeau, introduced legislation in 1980 which sought the 'Canadisation' of the petroleum industry. It gave substantial tax exemptions/grants and other advantages to enterprises with over 50 per cent of Canadian ownership, for the exploration and development of oil in offshore and Arctic areas. It also gave the state-owned energy company, Petro-Canada, the right to acquire up to 25 per cent interest in these areas. Later, in 1985, the concessions to Canadian companies were reduced, on the election to power of the Progressive–Conservative Party, led by Brian Mulroney.[28] Moreover, the Mulroney Government gave Petro-Canada a new mandate:

> to operate in a commercial, private-sector fashion, with emphasis on profitability and the need to maximise the return on the Government of Canada's investment. In this regard, Petro-Canada is not to be perceived in the future as an instrument in the pursuit of the Government's policy objectives.[29]

Japan, since its industrialisation, has been one of the economies most vulnerable to interruptions of essential flows such as energy. In 1973 Japan's energy imports represented 94 per cent of total energy consumption, with oil representing 75 per cent of the country's primary energy requirement, and the politically sensitive Middle East representing some 85 per cent of imported crude oil. Thanks to a vigorous strategy of reducing its energy dependency and vulnerability, these figures were reduced in 1984 to 85, 62 and 70 per cent, respectively.[30] One of the favourite policy instruments at the command of the Japanese Government was the Japan National Oil Corporation (JNOC), created in October 1967. Its main functions comprise:

— Providing equity capital and loans to the Japanese companies for petroleum exploration projects (including natural gas, oil, shale and oil sand).
— Direct financing to government or national oil companies of oil-producing nations for exploration and development projects.
— Providing guarantees for liabilities incurred in financing petroleum exploration and development projects overseas.
— Acquisition of rights for exploration for petroleum reserves overseas.
— Conducting gratuitous geological surveys for petroleum exploration projects upon the request of foreign governments.
— Technical assistance for exploration and development of petroleum resources.
— Research and development and the training of engineers.
— Leasing of the necessary machinery for petroleum exploration projects.
— Stockpiling of petroleum.
— Providing loans for oil purchases to build up stockpiles.
— Providing equity capital and loans needed for installation of the necessary facilities for stockpiling.
— Gathering information concerning the exploration, development and stockpiling of petroleum.[31]

To carry out its many functions, JNOC has been endowed with substantial financial resources (e.g. ¥1,028 billion in equity capital and loans for exploration by 1984) and has been favoured with fiscal incentives to facilitate exploration and development activities in Japan, and in authorised projects overseas. JNOC has also been entrusted with managing the Government's oil stockpile and aiding the private sector in meeting the required stockpiling targets (its equity capital and loans commitment reaching some ¥360 billion by 1984).[32]

After the Second World War, the US Government considered directly purchasing a controlling interest in a privately-held US oil concession in Saudi Arabia (Aramco) to boost its production, and the construction of a government pipeline from the Gulf to the Mediterranean. Direct US government participation was, however, opposed by US oil companies which did not have a stake in Aramco.[33]

US state ownership in the oil sector was advocated in mid-1979 by leaders of organised labour, the AFL-CIO, to prevent companies from behaving against the public interest. The United Automobile Workers, a union with 1.5 million members, went further. At a massive work stoppage (with a letter-writing campaign) on 22 August 1979, they called for, first, the setting-up of a national state-owned oil company; second, the control of domestic prices of oil and gas; and third, a government take-over of the importation of oil from all private companies. The latter refers to the proposal for a 'secret auction for import tickets' supported then by thirty American Senators. It was contended that this would lead to cutting import prices in the process of bidding for the tickets.[34] The UAW president, Douglas Frazer, gave his board's position as follows:

> We re-emphasise that mere efforts to strengthen the market mechanism will not work. Decision-making with respect to petroleum is not made by separate small entities, it is made by foreign nations and multinational corporations. That can only be counteracted by direct US Government activity in importing, developing and producing petroleum.[35]

The state-ownership call is not prompted solely by the fact that oil has assumed a 'public utility' role. Indeed, several public utilities in the United States—notably telecommunications, power generation, water distribution and other municipal services—are owned and run by the private sector. The advocacy of state ownership has generally been based on the fact that oil is a strategic commodity crucial to the well-being of society and the security of the United States. For the proponents of state intervention in the United States, it is presumed that Federal ownership would enhance the effectiveness of the administration's energy policy—not only at the level of information and logistics but also at the level of planning and operating various key activities of the oil industry. Such ownership, it is claimed, could severely curtail oil companies' ability to circumvent laws and regulations.[36]

The intellectual underpinning for state intervention in the US energy sector has been provided by academic circles in the United States. Various academics have called for:

- the creation of a state monopoly responsible for oil imports;
- the creation of a state corporation for investment and research in non-oil sources;
- the development of oil on federal land by a state corporation.[37]

This mode of thinking has not, however, won the support of any significant or influential section of the American general public, law-makers or academic circles for long.

Deeply concerned about disruptions in oil supplies following the Arab–Israeli war of October 1973, the British Labour Government passed the Petroleum and Submarine Pipelines Act, 1975, which led to the creation of the British National Oil Corporation (BNOC) on 1 January 1976. The motives behind the creation of BNOC were to give the United Kingdom a guaranteed access to its own oil (in the eventuality of shortages) and to have a reference price on which taxes could be levied. A subsidiary motive appears to have been to increase 'the information flow to the Department of Energy and other government departments', according to the first chairman and chief executive of BNOC, Lord Kearton. He affirmed that 'before BNOC came along, the Foreign Office depended very much on what Shell and BP chose to tell them'.[38] BNOC's role in this respect was to give government departments objective information, that is, not biased in favour of the majors.

Upon its establishment, BNOC obtained a large portion of crude oil produced from the British continental shelf through three main avenues:

— handling royalty oil accruing to the British Government through the functioning of leases awarded to exploration–production companies;
— exercising the right to buy up to 51 per cent of British North Sea oil at the official price;
— commercial purchases.

In the absence of a downstream operation, BNOC sold its crude oil in the open market.

There have been counter-movements to state interventionism. Certain governments have clamped down on the growth of state enterprises, favouring deregulation and the privatisation of various business activities hitherto under state control. In the oil sector, the British Conservative Government succeeded in cutting down the state's controlling interest (previously above 50 per cent) in British Petroleum to 31.7 per cent. It furthermore sold other state-owned petroleum enterprises to private interests—British Gas Onshore Oil Assets, Enterprise Oil and Britoil.[39]

In response to the programme of privatisation espoused by the British Conservative Government in the first half of the 1980s, BNOC's oil exploration and production activities were transferred in August 1982 to a new company, Britoil. The shares of the latter were subsequently sold to the public.

In March 1985 the British Government decided to abolish BNOC, prompted by losses incurred by the Corporation over the year 1984/85, reaching a peak of £49.8 million in the first quarter of 1985. Losses were due to the crude-oil trading activities of BNOC. The Corporation lifted 51 per cent of the oil produced from the UK continental shelf, whose production was close to 2.5 million barrels per day in 1984. BNOC obtained the oil at official term prices; about 14 per cent of the oil was resold at spot prices in the open market. This yielded BNOC profits prior to 1983, when spot prices were above official prices. The price situation changed over 1984/85 when spot prices fell by as much as $3 per barrel below official prices. The UK Government expected the weak price to continue when it decided in favour of abolishing BNOC, in accordance with a statement to Parliament by the minister of state for energy on 13 March 1985.[40]

Privatisation of state-owned enterprises (i.e. the transfer of assets to the private sector) is often linked to a policy of liberalisation; namely, the removal of barriers to competition. Besides the ideological objective of reducing the size of the public sector, privatisation/liberalisation is primarily sought to reduce or eliminate monopoly power and to enhance competitiveness, thereby promoting the efficient allocation of resources. In so far as the private discount rate is higher than the social discount rate (in the United Kingdom, these were estimated at 10–15 per cent versus 5 per cent respectively over the 1970s and 1980s), privatisation would accelerate depletion. On that score, privatisation, unless regulated, could act against the national interest.[41]

Evidence regarding the effectiveness of the state-enterprise approach *vis-à-vis* the private-enterprise approach in promoting national interests in the energy field is not conclusive. The policies and experience of individual countries have to be analysed in the context of each country's socio-political context, which could evolve considerably over the years. Transnational enterprises, in contrast to a national agency, have had to balance competing national interests, bearing in mind longer-term corporate interest. A transnational company's global interest can be at variance or even in conflict with that of a single country at a given period of time.

An ex-chief executive of a state-owned French group felt that government shareholding of oil enterprises had serious drawbacks. Chief among these were: belated reactions to market realities; divided, incoherent or shifting policies representing the different views of government ministers or departments; short-term 'political' perspectives aimed at protecting ailing economic groups to the detriment of the efficient and dynamic enterprises; and an authoritarian approach by the government to resolving differences with management.[42]

WINDFALLS AND RISKS

Ascertaining Normal Profits

Competitive market conditions should yield normal profits for the productive and efficient operator. Normal profits under competitive market conditions, according to established economic principles, represent a reward for *functions* undertaken or *sacrifices* made. The latter generally comprise:

— risking resources under conditions of uncertainty in order to launch, develop or expand a business activity;

— successful commercialisation of new products, services, technology or know-how;
— efficient management and logistics.

Profits in excess of the normal can be attributed to a number of factors, chief among which are: economic rent; monopolistic positions; windfalls. Economic rent can be defined as 'surplus profits' resulting from the ownership of an asset whose higher yield is due to its inherent richness or superior productivity, compared with other high-cost producing units which just manage to break even in a competitive market. This applies, *inter alia*, to agricultural land, mineral rights and tourist sights. In the case of subsoil riches, the landlord—with the notable exception of the United States—is generally a public authority. Profits resulting from monopolistic positions represent the ability of one seller or of a number of sellers to use their dominant bargaining position or their capacity to constrain trade to extract larger profit margins than would obtain under competitive conditions. Windfall profits represent unexpected profits which may result, *inter alia*, from disruptions in supplies or from sudden increases in demand. In 1979 the chairman of the US Council of Economic Advisers, Charles L. Shultze, defined windfall profits thus:

> Windfall profits are returns that accrue to the owner of an existing investment that arise from events outside his control and that are in excess of the level needed to replace and expand that investment. The owners of many existing oil properties made the decision to develop the properties on the basis of domestic—and world—oil prices that were a fraction of today's level. Their good fortune today results from actions in which they had no part—the world-wide short supply of petroleum, the actions of OPEC, the timing of the US decontrol decisions, etc. Thus, to a large degree, the increased revenues these producers will be receiving do indeed qualify as windfalls.[43]

Windfall profits are normally considered to be due to the government in so far as the latter is the representative of the national interest. The government is often responsible, directly or indirectly, for the creation of such profits. Examples of government actions which produce windfall profits are:

— administered prices set up, accepted or condoned by the authorities;
— protective trade barriers;
— monopoly positions created, facilitated or condoned by the public authorities of the country in which the enterprise operates;
— favourable fiscal treatment of an operator.

State recouping of windfall profits is advocated not only on the grounds that the enterprise has made no effort in their generation; it is also advocated on the grounds that the state can put these windfall profits to better use (i.e. for the public interest) than the enterprise itself. Put differently, the social return of recouped windfall profits is considered superior to the private return of the company or companies concerned.

Governments of oil-exporting or oil-importing countries have been trying over the years to recoup excess profits—whether resulting from windfall events, monopolistic positions or economic rent. Empirical studies suggest that 'the great majority of fiscal systems on oil exploitation are not well designed to capture a large share of economic rents to governments while at the same time ensuring that incentives to explore for and develop fields are unimpaired'.[44] Indeed, several small fields are left unexplored or undeveloped owing to fiscal disincentives, while most fiscal regimes are not progressive enough to capture the economic rent of large, low-cost oil-fields.

The principle of taxing excess profits has long been espoused by various public authorities, notably with respect to the petroleum industry. For a Canadian example, the British Columbia Energy Commission stated in 1973 that:

> The owner of a resource [the Province of British Columbia] obviously can only be properly compensated for it if the resource is fairly priced on the market and the surplus over and above a fair return to the producer is paid to the owner; it is the function of a royalty system to ensure that the owner receives this surplus.[45]

After deliberating on evidence produced by companies exploiting the natural gas of the province, and hearing other witnesses, the Commission concluded:

> While analysis suggests that some of this increase should go to the producers who are not receiving an adequate return at the present prices, it is clear that the provisional royalty should be revised so that the surplus generated by these windfall price increases will accrue to the Province.[46]

The above principle, aimed at recouping surplus or windfall profits by the state, as owner of natural resources, was also adopted by the Canadian Petroleum and Natural Gas Act of 1976. One finds similar approaches in Norway and the United Kingdom with respect to oil ventures in the North Sea. The rationale for the Norwegian Special Tax on petroleum was stated by the Finance Committee of the Norwegian Storting (Parliament), thus:

> The unusually sharp rise in petroleum prices, since the autumn of 1973, has created an entirely new profit situation, also for those companies producing petroleum in the North Sea. Even though the costs for oil and gas production have also risen sharply, production is far more profitable than it was on the basis of former crude-oil prices.

> The entirely new profit situation is the basis for the government's view that the taxation of oil companies should be altered so that a greater share of this additional revenue will acrue to Norwegian society than is the case according to existing rules.
>
> The majority points out that the petroleum resources on the Norwegian Continental Shelf are the collective property of the Norwegian people. In the view of the majority, Norway's desire to secure a substantial share of the values created on the Shelf is therefore not unreasonable.[47]

Similarly, a spokesman for the British Government made the following statement of policy with respect to excess profits:

> the original licensees are subsidiaries of foreign-owned companies and there could be a large balance of payments loss through remittance overseas of their share of excess profits unless corrective action was taken . . . In fact the only way of doing this is by increasing the government's share of the profits.[48]

The chief executive of a major transnational oil company admitted that the swelling of operational revenues between 1974 and 1980 had not been due to company efforts. He said, 'I don't think we or any of the oil companies deserve any great credit for increased revenues. They are generated by the price that OPEC puts on its crude oil'. He added that

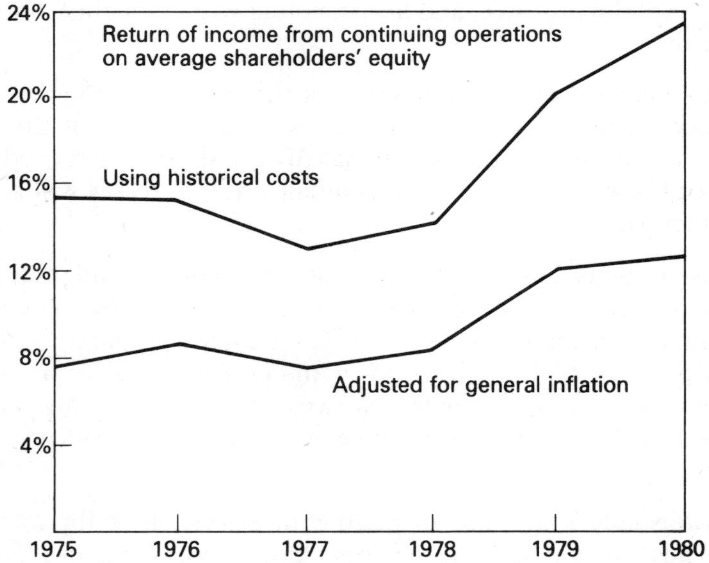

Figure 4. Measuring the inflation distortion in Exxon's profits.
Source: Judith Maxell, 'Exxon's Profits: More (or less) than Meets the Eye', *Esso European Review*, London, Winter 1980–81, p. 7.

much of these revenue increases had been paid in taxes, the companies acting as a 'tax collecting agency'.[49]

In periods of inflation, unadjusted figures could produce an illusion of excess profits. Thus, unadjusted rates of return on average shareholders' equity between 1975 and 1980 was about twice as high as the adjusted figures for general inflation for the world's largest petroleum company (Figure 4). The rates of return of private US petroleum companies in foreign countries have been substantially and consistently higher in developing countries as compared with developed countries.[50] In 1980–81 the figure for the former group was 49.9 per cent, as compared with 23.2 per cent for the latter. In 1982–83, the figures were 25.2 and 13.6, respectively.[51] The higher return obtained in developing countries can be attributed to any of the following important factors: premium for risks (political, geological, market); monopolistic profits; economic rent.

Risks of a political nature are not limited to developing countries: they also include Western industrial democracies. To illustrate, the world's largest oil company accused the French authorities of arbitrary and discriminatory fiscal charges on its exploration activities in the Paris basin.[52]

INTERVENTIONISM ABROAD

Government instruments used in controlling or breaking the economic 'power' of rivals or adversaries vary from the subtle and moderate to the direct and forceful: they stretch from diplomacy to military operations. The use of some of these measures is reviewed in the following, covering both transnational oil and non-oil countries.

Diplomacy and Economic Sanctions

The management of relations among nations is often conducted by negotiation between representatives of governments. This approach has often guided protagonists in industry, including the world oil industry. Private enterprises, concerned about protecting their interests in foreign countries or seeking entry into hitherto closed overseas markets, have solicited the support of their home governments.

This has been true of Western enterprises, which have tended to call on the active support of their home governments to obtain advantages or counter obstructions in host countries. The US Government's support for US enterprises overseas has been forthcoming during periods of perceived shortages of domestic energy supplies. This support has nevertheless remained less prominent in comparison with that of European nations. As one US academic put it, 'On the US side, the engagement of high government officials was much shallower. From the viewpoint of American oil companies, sympathetic support of the US government was a force to be held in reserve, not to be allowed too prominent a place at the negotiating table'.[53] European governments have been more involved in supporting companies; this was particularly the case with the British Government's support for British Petroleum and the French Government's support for Total-Compagnie Française des Pétroles in their encounters with host countries prior to the 1970s.

In 1982 one US source acknowledged the influence the US Government had over one major OPEC member: 'The United States uses its bilateral relations with producing countries, particularly Saudi Arabia, to support IEA decisions'.[54] In 1981 the Iraqi oil minister contended that US influence on OPEC's policies materialised through US influence on Saudi Arabia. He stated:

> During the [25–26 May 1981, Geneva] Conference, an American journalist asked me whether this single member country [Saudi Arabia] would accept a price of $35/b, I said I sincerely hoped so, but he said he would doubt it, and added that he would be surprised if this single country raised its price to $35. 'We in the American press are certain that this single member country will not accept a price above $34/b.' I said exactly the same thing at Bali Conference and pointed out that the US determines OPEC's policy through this single country which enjoys high production rates with a huge capacity [Saudi Arabia had about 40 per cent of OPEC's total exports in 1981].[55]

The above-mentioned desire to raise the price to $35 per barrel for marker crude was then expressed by eleven of the thirteen OPEC countries—Saudi Arabia objecting and swaying the United Arab Emirates to its side. The Saudi position of $34 per barrel was later adopted by OPEC ministers at their conference meeting of October 1981 at Geneva.

As a secondary approach to clipping 'OPEC power' by influence from within, industrial countries have sought to discourage potential new entrants from considering OPEC membership. The industrial countries' persuasive (or disuasive) approach can be achieved by various methods, such as:

- offering (or denying) the oil-exporting country willing (or refusing) to co-operate with them economic advantages;
- protecting (or destabilising) the economic/political system of the oil-exporting country concerned depending on its co-operation (or non-cooperation).

To induce co-operation, one economic instrument that has been used is trade. For example, the world's major industrial country and a large oil importer, i.e. the United States, has revoked the 'most-favoured-nation clause' for OPEC countries. The US Government has further encouraged non-OPEC oil producers. Thus, it gave Canadian and Mexican oil unrestricted access to US markets in its policy of Western Hemisphere collaboration and denied the UNCTAD-negotiated 'Generalised System of Preference' (GSP) to OPEC member countries and to similar producer organisations. This move was based on the assumption that such organisations are bent on raising prices or constraining supplies of vital commodities.[56]

In order to influence other governments or countries' elites, various parties have also resorted to covert action. This applies to totalitarian regimes as well as to liberal ones. Thus, the US Central Intelligence Agency (CIA) has used transnational companies to install its agents, disguised as corporate executives. One such company, Chase Manhattan Bank, reputedly stopped accepting CIA agents, fearing that disclosures could hurt the bank's business.[57]

Short of threats and the eventual resort to political destabilisation or belligerency, states have used economic sanctions as a means of pressuring foreign parties into compliance with their wishes. Economic sanctions comprise a whole array of measures. They are illustrated below by reference to the freezing of financial assets and the disruption of trade.

Financial assets

The strategy of freezing financial assets was espoused by the US Government in a special case regarding Iran in mid-November 1979. The United States was provoked by the seizure of their embassy in Tehran and the holding as hostages of the US embassy staff in Tehran by militants supported by the Iranian authorities. The latter were prompted by the US refusal to extradite the deposed Shah, who was undergoing medical treatment at the time in the United States.

Further exasperated by a press conference of the then-acting foreign minister, Abol-Hassan Banisadr (in 1980–81 Iran's president), who expressed his government's intention to withdraw its deposits from US banks, President Carter acted under the International Emergency Economic Powers Act of 1977 and the National Emergencies Act, blocking:

> all property and interests in the property of the government of Iran, its instrumentalities and controlled entities and the Central Bank of Iran which are or become subject to the jurisdiction of the United States or which are in or come within the possession or control of persons subject to the jurisdiction of the United States.[58]

That jurisdiction was interpreted as covering not only US-based financial institutions but also the foreign branches or affiliates of American financial intermediaries operating outside the United States.

The President's action was specifically designed 'to protect US claims' on Iran in the case of withdrawal of Iranian funds. The official Iranian assets held in the United States and with the overseas branches of US banks are given in Table 17.

Table 17 Estimates of Iranian assets blocked on 14 November 1979 ($m.)

	Estimated as of 19/11/79	Census as of 25/10/80
US government securities	1,235	1,274
Treasury bills and certificates	1,234	n.a.
Treasury bonds and notes	1	n.a.
Federal agency issues	n.a.	n.a.
Corporate bonds Corporate stocks	35	65
Commercial banks liab. in US	1,426	1,177
Demand deposits	n.a.	807
Time deposits	n.a.	150
Other	n.a.	219
Subtotal	2,696	2,516
Non-banks' liabilities	553	1,463
Direct investment	0	n.a.
US government accounts	435	472
Total investment in US	3,684	4,451
Gold held in US	638	636
Assets blocked abroad		
Banks	4,523	4,546
Non-banks	n.a.	63
Total	8,845	9,696

Source: US House of Representatives—Committee on Government Operations, *Federal Response to OPEC Country Investments in the United States* (Part 1, Overview), 97th Congress, 1st Session, Washington DC, 22–23 September 1981, p. 176.

It is pertinent to note that the impact of shifting US dollar deposits to non-American banks and/or to other currencies could be readily neutralised by appropriate swap-and-loan arrangements in the international banking community. The international financial markets have a remarkable capacity to make necessary adjustments in managing international liquidity. A bank bears risks in having concentrated deposits, since the withdrawal by a large depositor could involve, to avoid illiquidity, a recourse to the inter-bank market at premium interest rates; the profit margins of the bank will suffer as a result. The analysis of a US bank official confirms this interpretation:

> A popular misconception views such a withdrawal as analogous to a run on a bank by depositors, having all the deflationary effects associated with hoarding. This view, which became prominent again

during the threat of Iranian withdrawal of funds from United States banks, is misleading and exaggerates the costs of deposit transfer. The withdrawn funds are not hoarded but are redeposited in other Euromarket banks that are then in a position to supply funds to the institutions suffering the original withdrawals. As a result, the original banks exchange direct deposits for inter-bank borrowings. This, however, is not without costs; the original institutions may have to pay a premium to raise an extraordinary amount of funds in the inter-bank market and their profits may fall as a result. So some risk attaches to concentrated deposit holdings by country as well as to concentrated claims positions.[59]

Following the US freeze of Iran's official assets, several US banks have striven to secure protection for their claims or the claims of their clients on Iranian assets. It is interesting to note in this respect the action brought by an American bank, Morgan Guaranty Trust, which was generally thought to have more deposits from official Iranian sources than loans. Morgan sought and obtained a distraint order against Iranian assets, notably Iran's 25.01 per cent holding in Krupp Hüttenwerke of Germany.[60] As a result, 'German officials have expressed distress at US banks waging economic war on Iran within their territory'.[61]

Other US banks (notably Chase) were eager to declare Iran in default following these banks' inability to make transfers from Iranian frozen assets, and before allowing Iran time to make alternative arrangements for debt settlement. According to one trade source, 'The very deep resentment harboured by bankers outside America against the US administration and the major US banks will not be dispelled easily'.[62]

Commenting on the use of the asset freezing weapon, an official of a capital-exporting country (the Kuwait minister of finance in 1979, Abd al-Rahman al-'Atigi) observed that this strategy 'epitomises the principle of the use of force'.[63] He added:

> I would just like to repeat what I recently said to Mr Miller [then the US treasury secretary], namely that the words of comfort now being conveyed to us by the United States are really of no use. The West tells us that the assets of the Gulf countries are safe. But once confidence is shaken, it is very difficult to restore it. And media campaigns in the US do not generally distinguish between Iranians and Arabs. They just lump them all together as oil producers and petro-dollar holders and call for them to get out of the US. How can you restore confidence in these circumstances?
>
> Perhaps the US would have had an excuse for retaliating in kind against the Iranians by holding their ambassador or something like that—but not for freezing their money. It is not just a question for Iran, the whole world is shocked by this freeze which has general implications of a most dangerous and serious kind. The most serious of these

> implications concerns the interpretation of national security. Which is to be the authority to decide about national security? Each country has a very vague definition of this. It seems that anyone can repudiate debts and freeze assets on the grounds of national security. If the US with its pre-eminent position in the world can do this, what about everyone else? This is really something shocking and, as I said before, it will be difficult to restore confidence.[64]

The Saudi minister of finance considered the US freezing of Iranian assets 'very serious'. It has prompted the Saudis to seek further diversification of their assets, and additional legal protection.[65] The comment by an official US source reflects on the international implications of such a freeze: 'The act of freezing Iranian assets by the US government in foreign branches of US banks operating under the jurisdiction of their host countries introduced new uncertainties that may continue to exercise influence on international financial flows for years to come'.[66]

On 3 April 1982 the British authorities resorted to freezing Argentina's financial assets and to denying it new sources of funding from London's financial market. This punitive action was taken in the wake of the Argentine invasion of the Falkland Islands.[67] The Bank of England's deputy governor, Christopher McMahon, told foreign bankers in London on 26 May 1982 that the financial sanctions against Argentina were 'wholly effective'. In the first month of the sanctions, £2.2 billion of foreign currency deposits were withdrawn from London, in favour of New York and Zurich. Nevertheless, with the moderate interpretation of rules governing the British sanctions against Argentina, the confidence of foreign depositors has not been lastingly shaken.[68]

The strategy of freezing foreign assets has not only been used by industrial countries, such as the United States, the United Kingdom and France, at different points in recent history. This economic weapon has also been used by developing countries. Thus, Egypt resorted to the same strategy against its fellow Arab countries by freezing $2 billion of official funds deposited by Kuwait, Saudi Arabia, Libya and Iraq in its banks. The reason advanced by the Egyptian authorities was the protection of their national economy against the potential damages of an untimely withdrawal of these deposits.[69]

Trade

Trade sanctions have stretched from denial of favourable treatment in trade to total embargo or boycott. The latter can apply to a wide variety of goods and services, such as food and energy, high-technology equipment, credits, transport facilities, or less important products. The impact of these measures can be assessed from different time perspectives. In the short term, the disruptive consequences to industry and consumers, and the adverse psychological impact, are likely to be considerable. In the longer term, alternative supply/outlet routes or substitute products and services can be developed.

Complex factors must be assessed in the choice of measures. The ultimate objective is to weaken the adversary. Effectiveness can be measured according to:

— the political resolve and the capacity of the initiator to inflict comparatively greater harm on the adversary than on one's self and one's allies;
— the inability of the opponent to sustain for long the sacrifices implied by such sanctions, and the imminent capitulation to the author of the sanctions.

Embargoes and boycotts have often been seen as the ultimate economic weapon in a state's arsenal against its opponents. In the case of the United States, cereals and high-technology goods for the Soviet Union were subject to embargo. In 1981 and 1982 President Reagan placed an embargo on the sale of equipment and technology for the construction of a Soviet Trans-European natural gas pipeline—following the imposition of martial law in Poland, thought by the US Government to be a result of Soviet pressure.

Embargoes have also been implemented against individual OPEC countries. Following the hostage crisis, Iran was subjected, in April 1980, to an embargo on practically all shipments of US goods. Food and medicine were restricted to minimal amounts and they had already been brought to negligible levels by the refusal in November 1979 of US dock workers to handle shipments destined for Iran.

Another OPEC country, Libya, was subjected, from March 1982, to a ban on US exports of oil and gas technology, and of 'items controlled for national security purposes', as well as to a ban on imports into the United States of Libyan crude oil. These sanctions were prompted by alleged Libyan support for 'acts of international terrorism'. The US Administration advised its citizens living in Libya to leave that country for reasons of safety—an anxiety reportedly not shared by the persons concerned.[70]

US cereal exports to the Soviet Union and other countries were subject to embargo on more than one occasion. Notwithstanding the pledge in his 1976 presidential campaign 'never to embargo US grain shipments because of the harmful effects on US producers and foreign countries . . . he [President Jimmy Carter] has ordered the most massive cancellation of grain contracts in history'.[71] The embargo was imposed against the Soviet Union early in 1980 in reaction to the latter's military intervention in Afghanistan. Referring to the cutbacks in US exports of wheat to the Soviet Union, one Western analyst commented:

> At one time breaking sales commitments was mainly confined to politically unstable 'banana' republics. Now it has been done by the

> richest, and most powerful, country in the world. This has raised doubts among potential buyers about the US as a reliable supplier. It is one thing to cut soya bean exports, as happened in 1973, because of a poor crop, but quite another to cite a political motive, whatever it might be . . . Any country will think twice about relying on US supplies, whatever inducements may be offered.[72]

One official US source pointed out a further serious limitation of US grain embargoes in so far as they put the onus of US foreign policy on a single sector of the national economy:

> The embargo was widely criticised from its inception by the US farming industry as inflicting a disproportionate burden of a US foreign policy action on US farmers. The criticism intensified as many analysts began to question the effectiveness of the embargo in the light of increased shipments by other nations.[73]

Some other US officials went to the other extreme, in favouring unrestricted grain sales to the Soviet Union. In so doing, they argue, the United States hooks the Soviets into dependence on the West in a strategic area, namely food. Such dependence should lead to a depletion of Soviet foreign exchange reserves; as a result, it decreases that country's capacity to buy Western technology in order to raise its productivity in civilian sectors of the economy, and eventually attenuates its defence capability.[74] While, on the one hand, selling grain to the Soviets depletes their foreign exchange, some US officials have argued, on the other hand, that selling sophisticated equipment used in the Soviet gas industry raises Western Europe's dependence on Soviet energy, provides the Soviet Union with needed foreign exchange and enhances the technological know-how of the Soviet Union, which could be used for military purposes.[75]

Would the embargo weapon be more effective in the case of advanced technology than in the case of food products? To illustrate, President Carter instituted by executive order early in 1980 an embargo on the sale of computers to the Soviet Union as a protest against the latter's invasion of Afghanistan. President Reagan, also by executive order, decided in December 1981 to impose a ban on goods manufactured in the United States being used in laying the Soviet gas pipeline to Western European countries. In June 1982 he extended the ban world-wide to forbid the use of advanced industrial equipment that employed US technology or which was manufactured under US licence. This ban affected US firms' subsidiaries, affiliates, joint ventures and associates overseas (the latter were defined to include as low as a 10 per cent US interest in their equity), as well as foreign companies manufacturing under US licence. President Reagan's aim was to prod the Soviet Union to use its influence in getting the Polish Government to lift martial law restrictions.

Various US observers reflected on the unproductivity of such an embargo. One executive commented, 'Using trade as a weapon or a lever only works in the case where you have a consensus of countries doing the embargoing. The US no longer has a lock on technology'.[76] Thus, to replace the embargoed US computers, the French group CII Honeywell–Bull filled in the gap.[77] Moreover, enterprises from the country imposing the embargo (here the United States) have been tempted into finding ways and means to circumvent export controls—especially when the rewards are enticing.

That US grain embargoes proved of limited effectiveness, despite the US control of some 50 per cent of world exports, was due to three main reasons:

— The embargo was not comprehensive in so far as it applied only to quantities promised in excess of an agreed minimum (8 million tons).
— Western allies of the United States—notably Argentina, Australia, Canada and the European Community—failed to abide by US admonitions to join its embargo.
— The target nation(s) could not be isolated. Swapping of customers and supplies, diversions and trans-shipments can thwart an embargo, especially when placed on commodities which can be produced or trans-shipped by different countries.

These constraints apply equally to a range of commodities.

Developing countries have also brandished the 'economic sanctions' weapon against other countries, developing or developed. This weapon has in general been even less effective in achieving their objectives, as compared with the industrially developed nations. Indeed, developing countries have limited bargaining power, largely owing to their weaker and less versatile economic basis, as well as to their limited military leverage. Notable examples of economic sanctions imposed by developing countries for political reasons are:

— an Arab economic boycott of Israel and companies doing business in that country;
— a short-lived ban on Arab oil exports late in 1973 to selected destinations, mostly aimed at the United States, which frustrated Egyptian and Syrian military activity directed at recovering their territories occupied by Israel since the hostilities of 1967;
— a ban by Nigeria and other African–Arab states on oil exports to South Africa to protest against apartheid legislation.[78]

Oil-export bans have generally proved to be of limited or no effect. Indeed, Arab oil-exporting countries later deemed it counter-productive to the achievement of their national aspirations; they thus rejected its use at their mid-1982 OAPEC ministerial meeting when considering moves to press the United States to use its influence to obtain the withdrawal of Israeli occupation forces from Southern Lebanon.[79]

Embargoes can also exist among friendly countries, despite their membership of the same political/military alliance. Embargoes on exports of soya beans and derivatives were used in this way by the United States in 1973, despite firm contracts entered into by European and Japanese importers with their US counterparts. The US Government followed a policy aimed at moderating domestic price increases in the grain and animal feed market, following a poor crop. But a country such as Japan, highly dependent on soya as a staple consumer product, found the US Government action particularly damaging to the Japanese consumers' well-being.[80] As another illustration, one could cite President Reagan's embargo on imports of iron and steel piping from the European Community in the latter part of 1984.[81]

Economic boycotts in an interdependent world can produce paradoxical situations. One illustration of this is Kuwait's boycott of Trans-World Airlines in February 1981 as a result of the airline's advertisement showing al-Aqsa mosque in Arab East Jerusalem as part of the Israel capital. The airline flies regularly to Tel-Aviv and has sales offices in Kuwait. It manages, through an affiliate, the Hilton hotels in Israel, Kuwait, Abu Dhabi and many other Arab countries. Furthermore, investors from Kuwait and Abu Dhabi, according to company officials, reputedly held shares in the airline.[82] Arab boycotts of Israel have invited forceful reactions by a number of Western countries, notably the United States, in the form of heavy penalties to US companies which submit to such boycotts.[83]

Extraterritoriality

The impact of enforcing the application of national legislations and sanctions on other countries can undermine transnational co-operation. This interpretation is generally shared by government and business analysts, as poignantly put by a US trade publication:

> The United States can try to be the world's policeman. But in the process of doing so, it will lose its friends, hurt the sick European economies, encourage an already vicious circle of economic retaliation and protectionism and, if it's lucky, inflict a pinch on the Soviet Union.[84]

In the judgement of an official US publication, 'the economic pressure on the Soviets was marginal, while the political ill-will that the [1981–82 US] embargo generated on the part of Western European allies and its detrimental economic impact on US firms were considerable'.[85] Indeed, outside the adversaries who were directly involved, economic sanctions have been a source of dispute among governments with respect to their enforcement outside their country of origin. Thus, the French Government ordered companies operating under French jurisdiction and producing equipment for the Soviet gas pipeline to refuse to comply with

US sanction orders—whether these companies were affiliates of US companies (notably Dresser and Rockwell) or French companies using US technology. The latter French groups were Creusôt–Loire (a licensee of Cooper)[86] and Alsthom–Atlantique (a licensee of General Electric). The British Government also formally declared its determination, on 30 June 1982, to protect British firms (notably John Brown, a licensee of General Electric of the United States) from compliance with American embargo decisions over the Soviet gas pipeline. Between 1980 and 1982 Japanese enterprises were also faced with a similar embargo on the use of US oil and gas technology in a joint venture with the Soviet Union to drill for oil off the Sakhalin Islands, notwithstanding the fact that Japan had signed an agreement to obtain a share of the oil and gas to be discovered and produced in that region.[87]

Various Western governments which are allies of the United States (notably the French, British, Canadian, German, Japanese, Italian and Dutch) and their parliaments or courts opposed the US policy of exporting its laws.[88] In particular, the opposition was to the policy of extending the application of US economic sanctions to cover US foreign affiliate, subsidiary, associate or licensee companies. Companies which did not comply were faced with stiff fines and the withdrawal of rights to buy US goods or know-how. Most governments considered the extraterritorial application of foreign laws and regulations to be an unwarranted and intolerable infringement of their sovereignty. European governments, furthermore, decried, in 1982, the double standards of behaviour and the expedient approach of the US Government bent on winning votes in an electoral year. In particular, the allies of the United States resented the US embargo on the sale to the Soviet Union of European equipment using US know-how for the construction of a Soviet gas pipeline to Western Europe, at a time when the United States had renewed its agreement with the Soviet Union for the sale of US cereals. The minimum guaranteed quantity of 8 million tons of cereal to be purchased by the Soviet Union in 1982 was exceeded (Figure 5). A British official commented thus:

> We don't like the practical application of the blockade, which damages Western European companies. It seems to us that Western Europe is being asked to pick up the bill for those measures while the American Mid-western farmer is allowed free and unrestricted right to trade with the Russians. That does not strike us as a fair distribution of the burden.[89]

The Western European governments' objections to the US application of sanctions can be summarised in four points:

- the assertion of extraterritorial rights that no foreign government has ever asserted in the United States;[90]
- the application of US sanctions retrospectively;

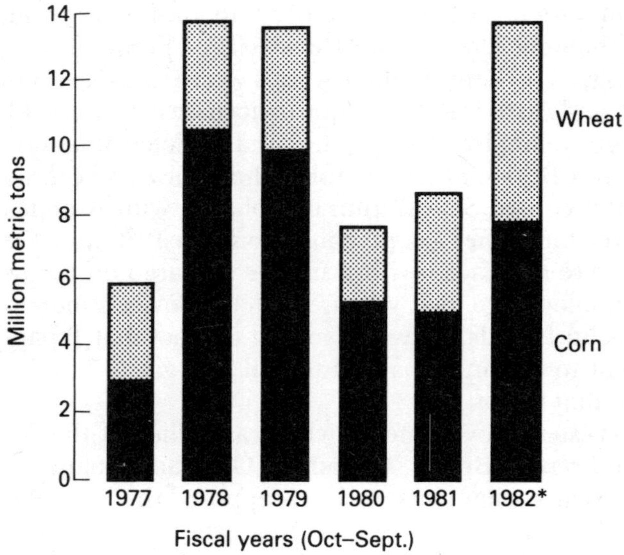

Figure 5. Soviet imports of US grains 1977–82.*
Source: Federal Reserve Bank of Chicago, *Agricultural Letter*, 25 June 1983, p. 3.
* Sales and exports as of 3 June 1982 for delivery this fiscal year.

— the absence of compensation for West European firms which had reached an advanced stage in the production of goods contracted, goods which were subject to damages for breach of contract—thus threatening companies with bankruptcy, especially those companies heavily dependent on Soviet orders;

— the absence of symmetry or parity in the application of sanction measures to various economic sectors and countries in order to balance out the adverse effects evenly among Western economies.

It was also recognised—by a congressional committee—that 'through US multinational firms and the accompanying extraterritorial applications of US law, the US government has created international tensions'.[91] The extraterritoriality also hurts US enterprises abroad. Commenting generally on the use of economic measures in foreign relations, one American executive reflected that:

> Laws and regulations are very clumsy instruments of foreign policy. They are blunt and inflexible. When our government attempts to wield them, resentment among those who are affected is only natural. The victims in the clash of sovereignties are often the multinational companies and their employees. American multinationals, especially, are in a vulnerable position. If corporations are perceived as the instruments of US moral or legal clout, they become the targets for those who would retaliate against this intrusion into their affairs. A large part of the diatribe against US multinational corporations is no doubt due to this reaction.[92]

Military Options

To reduce their vulnerability, resource-deficient countries have sought remedial action: this varies from, on the one hand, accumulation of strategic stockpiles to, on the other, the control of adequate foreign sources of supply. Foreign sourcing can be negotiated and contracted by commercial enterprises. Alternatively, the 'forceful' approach can appear attractive if the initiator has the material means and the predisposition to enforce envisaged action and if he expects a net pay-off from such intervention.

With some 50 per cent of commercial world oil reserves located in the Gulf countries since the 1950s, that region has a high international geopolitical significance, especially for major oil importers (which happen to be the developed countries). This significance, furthermore, has contributed to the heightened rivalry between the superpowers. Although the United States and the Soviet Union are the world's two largest oil producers, the first has become, since the 1950s, a major oil importer, and the second is finding increasing difficulty in satisfying its oil requirements and those of its East European allies. Yet the development of new oil-fields in Alaska (for the United States) and Siberia (for the Soviet Union) requires heavy capital investment. Some observers even surmise that the Soviet Union itself could well become a net oil importer by the turn of the century.

The media have given wide coverage to the possibility of military intervention by industrial powers in oil-exporting countries—a so-called 'crisis' approach to controlling the flow of oil supplies in terms acceptable to these powers. Special expeditionary corps, specially trained to intervene in the oil-exporting countries of the Gulf, have been set up by the United States and France. The Soviet Union intervened in Afghanistan with some 80,000 troops early in 1980, thus extending its reach to within 200 miles of the Gulf shores. To the Saudi minister for petroleum and mineral resources, 'Oil is the main reason why the Russians are penetrating and encircling the oil region', not only by their presence in Afghanistan but also in South Yemen and Ethiopia.[93] That oil is the main cause for Soviet involvement in the Middle East appears to be a simplification of a complex situation. The strategic geopolitical significance of the region has, for a long period of history, been at the centre of Russian, and later of Soviet, interest.

The military preparedness of Western powers, therefore, is not solely prompted by the desire to dissuade OPEC countries from espousing nationalistic or radical oil policies. This preparedness is also motivated by the necessity for fending off any possible direct or indirect action by the Soviet Union, which has had hegemonistic designs over Middle Eastern countries. In the 1980s the United States in particular obtained military facilities in five Arab countries to counter Soviet moves: Bahrain, Egypt,

Oman, Somalia and Sudan. It built up a Rapid Deployment Force (whose cost was estimated by US defence officials at over $2.3 billion) to protect the unimpeded flow of oil supplies from the Gulf to the West.[94]

The Soviet Union, for its part, warned against Western military intervention, even before the Iran–US crisis over the US embassy staff taken hostage and a few weeks before its own intervention in Afghanistan. President Brezhnev reputedly declared on 24 October 1979:

> Those who believe they will succeed with the help of a 'quick intervention force'—this new purely imperialistic creation—to lead Arab countries to renounce their sovereign right of controlling their own natural resources are mistaken. Wisdom is not reacting quickly at the trigger of a gun, but to react as necessary. The recurrence of imperialistic gunboat politics in its modern form can only create new dangerous focal points of tension, aggravate those already existing and worsen the international situation. Attempts to revive the inglorious policy of colonisers are doomed to failure, providing the people remain vigilant and resist them strenuously enough and in time.[95]

President Brezhnev, on the occasion of a state visit to India in December 1980, proposed that foreign military intervention in Gulf countries' affairs should be avoided. The Saudi minister for foreign affairs, Prince Saud Al-Faisal, commented that:

> Brezhnev has come out with the proposal of non-intervention by big powers in the Gulf region at a time when Soviet troops were still occupying Afghanistan . . . The proposal indicates only double standards in the Soviet approach. If the Soviet Union is sincere in its proposal, it should first withdraw its troops from Afghanistan.[96]

Various scenarios have been formulated for intervention by force. To some, they appear readily effective. Yet they bear important direct and indirect costs. It has generally been agreed that these instruments are best used dissuasively, as threats. Once applied, it is feared they could prove counter-productive since they could invite retaliation, and the process could spiral. Availability of energy supplies could, as a result of using coercive instruments, be reduced and consumers and producers would all be worse off.[97] Indeed, 'interventionist' strategies have prompted contingent 'confrontationist' strategies on the part of the oil-exporting countries. One measure that is contemplated is the sabotage of a few key production installations.[98]

Costs are likely to be material and moral. Besides the loss of oil property and supplies through fighting and sabotage, other costs include the absence of international legitimacy for an occupation force, and the loss of life that would be sustained by the invading force, perhaps beyond what public opinion in the interventionist country could tolerate. The military option can be seen as the ultimate weapon in the arsenal of confrontation. Short of fully-fledged direct military action by foreign powers, military operations could be carried out by proxy through local client states.[99]

Leaders of the Gulf countries are wary of superpower rivalry in their region. Even such a conservative statesman as the UAE President, Shaikh Zayed, reportedly considered that:

> the presence of US forces in the area is not something we expect from a friendly country like the US. The peoples of the world would co-operate and live in peace if it were not for the intervention of the big powers, so these powers must work to keep the Gulf region clear of their rivalry.[100]

The UAE minister for petroleum and mineral resources stated frankly: 'If we [exporters and importers] do not have an oil SALT [Strategic Arms Limitation Treaty] here, we might be faced with another World War on oil'.[101]

US LEADERSHIP AND GROUP ACTION

The United States can assume an important leadership role in providing an energy approach for the benefit of other oil-importing countries. This stems largely from the fact that it is the world's largest economy (judged by its GNP) and has been since the 1960s one of the largest customers and importers of oil in the world. Its options are in principle much wider than those of other oil-importing countries owing to its vast fossil-fuel reserves, its advanced technology and its financial and human resources. Its large and advanced economic structure and its currency (used as a reserve currency) enhance its leadership role in international affairs.

Large price increases initiated in 1973/4 by a few OPEC countries have led to a closing of ranks among major oil importers, namely the developed Western industrial states. Under the aegis of the United States, they formed the International Energy Agency (IEA) in 1974. IEA's founders saw OPEC as a challenge and have sought to reduce their dependence on oil from OPEC countries. A French minister for energy openly declared that IEA was created in 1974 as an instrument with which to confront OPEC. For this reason, he said, France did not join the IEA.[102] This interpretation of IEA's *raison d'être* is shared by various protagonists. Indeed, according to the leading architect of the IEA, US secretary of state Henry Kissinger, the creation of this consumers' organisation 'was to promote the cohesion of the industrial democracies in the field of energy, which in turn made a major contribution to improving the bargaining position of the consumers', as he euphemistically put it.[103]

IEA's architects considered collective solidarity essential for overcoming upheavals in oil supplies and prices. Individual, isolated, dispersed or inconsistent national action is often inadequate in reducing vulnerability to outside forces. Accordingly, the primary role of IEA has been that of averting or managing crises by policies of conservation, mandatory emergency stocks (ninety days of the previous year's net imports), demand reduction, sharing of supplies and joint research into the development of substitutes.[104]

IEA's role in the management of crises resulting from supply disruptions has yet to be tested. A government agency of one founding member stated that 'the United States was the moving force in establishing the IEA, and since its formation the IEA has been the centrepiece of US efforts to deal with oil at the international level'. Notwithstanding the importance assigned to concerted collective action, that same agency outlined late in 1981 the shortcomings of US policies in meeting IEA's collective objectives:

> [Its detailed] review indicates that in practical terms the IEA multilateral emergency programmes are characterised by various problems and are not strong enough to deal with the full range of disruption contingencies, and that the US has not effectively integrated its domestic and international contingency planning.[105]

A dilemma faces oil-importing countries: should they first 'drain' whatever domestic reserves they have and those of friendly non-OPEC countries, or should they keep these reserves for crisis periods and first drain those of OPEC countries? The drawback to the second approach is obvious: it would increase the power of OPEC countries in the short and medium term through the higher level of dependence of oil-importing countries on OPEC supplies, while the first approach could postpone that dependence to later years.

The dilemma is more acute when it is recognised that non-OPEC oil exporters and members of IEA are unwilling to sell to their fellow members at favourable prices. Indeed, a British secretary for energy frankly declared early in 1980 that it would be 'silly economics' to sell British oil at the official Saudi price level, which happened then to be significantly below the open market rate.[106] One problem with the policy of limiting 'drainage' is that no one knows for sure the size of reserves in the subsoil to be drained. Moreover, once commercial enterprises discover and develop oil reserves, they need to produce them to provide cash for amortising investments, remunerating capital and financing the discovery of new oil reserves.

Although France is not formally a member of IEA, it is indirectly represented through the participation of the Commission of the European Economic Community (EEC), of which it is a founding member. The EEC has, since its establishment in 1958, gradually evolved a set of supranational laws and regulations in various fields, including that of energy. Energy measures cover collective aid or group co-operation aimed at the conservation of energy, the diversification of energy supplies and the development of substitutes for imported petroleum. Not only does the EEC's energy policy aim at reducing vulnerability to price hikes and to restrictions and interruptions of foreign energy supplies; it also serves to take full advantage of economies of scale in projects which cannot be carried out efficiently by individual countries or their enterprises. EEC energy policy also aims at inducing member countries to avoid cost duplication in the field of research or investment. Its attempts at rationalising certain energy sectors can be illustrated by reference to coal-mining and to petroleum refining.

The EEC's Commission has obtained the agreement of the Community's energy ministers to phase out subsidies to high-cost collieries and to encourage companies to shut down obsolete oil-refining capacity.[107] In view of the contraction of European oil demand in the 1980s and the Commission's estimate of moderate future increases in that demand, consolidated data on existing and projected refining capacity serve to indicate the extent of excess capacity. This information is monitored by the Commission for the benefit of member governments and enterprises, and serves to assist the latter in readjusting their investment plans to harmonise capacity with expected demand, in order to avoid a further aggravation of the slump in the oil-refining business and the adverse impact on corporate earnings.

The Commission has also favoured the vertical integration process, whether of the refining/marketing companies moving 'upstream' or the oil-rich companies moving 'downstream'. It has also welcomed link-ups between crude-oil producers and refiner/marketers. This was notably the case in the joint refining/petrochemical venture in the Ruhr, whereby Venezuela's state-owned Petroven and West Germany's Veba Oel pooled Venezeulan crude-oil supplies and German know-how and market outlets.[108]

Group action, whether among oil corporations (see Chapter 1), exporter governments (see Chapter 3), or importer governments, is often prompted by challenges encountered or opportunities perceived. Thus, in the wake of the Great Depression of the 1930s, major oil companies sought agreement on stabilisation of market shares to *avert* the drastic impact of rivalry on corporate profits. In 1960 OPEC was created to *protect* its exports' proceeds against price reductions initiated by an oligopoly of transnational companies holding large oil-producing concessions in these countries. In 1974 major industrial oil-importing countries created IEA to help *prevent* excessive price increases and constraints on transnational oil trade. The dialectic of challenge and response can assist us in our understanding of changes in the structure and behaviour of protagonists in the world oil industry.

NOTES

1. See *International Herald Tribune* (*IHT*), Paris, 27 September 1979, p. 1.
2. For example, the US Under-Secretary for Monetary Affairs, Anthony Solomon, quoted in *Middle East Economic Digest* (*MEED*), London, 9 November 1979, p. 23.
3. A. L. Danielsen, 'Oil Price Regulation in the United States', *Journal of World Trade Law*, Geneva, September–October 1979, p. 392.
4. See, for example, US General Accounting Office, *Oil Supply Disruptions: Their Price and Economic Effects*, Washington DC, 20 May 1983; and Stephen P. A. Brown, 'Reducing US Vulnerability to World Oil Supply Disruptions', *Economic Review*, Federal Reserve Bank of Dallas, May 1982, pp. 1–13.
5. See, for example, 'US Pays less for OPEC Crude than that of "Friendly Countries"', reproducing *Arab News* article of 28 July 1979 in *OPEC Bulletin*, Vienna, 13 August 1979, p. 7.
6. National Coal Board, *Report and Accounts 1983/4*, London, pp. 11 and 68.
7. The US General Accounting Office, *The Synthetic Fuels Corporation's Progress in Aiding Synthetic Fuels Development*, Washington DC, 11 July 1984, p. 2.
8. Ibid., pp. 2–4; the US General Accounting Office, *Plans to Award Additional Financial Assistance to the Union Oil Company Oil Shale Program*, Washington DC, 23 July 1984, p. 11; and 'Pork barrel—or foresight', *Forbes*, New York, 20 May 1985, p. 108.
9. See, for example, 'Accord Nucléaire avec le Canada', *24 heures*, Lausanne, 25–26 May 1980, p. 7; 'Le Canada n'a pas encore levé l'embargo sur ses exportations d'uranium', *Tribune de Genève*, 23 August 1979, p. 5; and 'Uranium: le couteau suisse par le manche', *La Suisse*, Geneva, 29 June 1983, p. 43.
10. 'Nucléaire: Washington veut "punir" la Suisse', *Journal de Genève*, 23 Septembre 1980; and 'Swiss Nuclear Requests Blocked by US', *IHT*, Paris, 23 September 1980, pp. 1 and 2.
11. See 'L'Europe et le monopole américain', *Le Monde*, Paris, 7 May 1977.
12. J. H. Birnbaum, 'New Approach is Increasing Gasohol Sales', *Wall Street Journal*, New York, 7 September 1982, pp. 25 and 28; and US General Accounting Office, *Report to the Chairman, Subcommittee on Fossil and Synthetic Fuels, Committee on Energy and Commerce, House of Representatives*, Washington DC, 3 May 1985.
13. 'Oil Prices: The Medium-Term Risks', *OECD Observer*, Paris, July 1982, p. 8.
14. See *Petroleum Economist*, London, March 1980, pp. 99 and 101.
15. See 'Development of North Sea field delayed two years', *The Times*, London, 19 December 1980, p. 17; and 'Shell warns on oil output curbs', *Financial Times*, London, 27 September 1980, p. 3.
16. 'United States—Strategic Reserve Filling Completed', *Petroleum Economist*, London, October 1985, p. 370.
17. John Edwards, Commodities Editor, 'Counting the Cost of Strategic Stockpiling', *Financial Times*, 15 February 1983, p. 27.
18. *Oil Supply Disruptions*, op. cit., p. i.
19. International Energy Agency, *Energy Policies and Programmes of IEA Countries, 1984 Review*, Paris, OECD, 1985, p. 17; and US General Accounting Office, *Evaluation of the Department of Energy's Plan to Sell Oil From the Strategic Petroleum Reserve*, Washington DC, 5 June 1985, p. 56.
20. George Horwich and David Leo Weimer, *Oil Price Shocks, Market Response, and Contingency Planning*, Washington DC, American Enterprise Institute, 1984, p. 104.
21. US Department of Energy, *Annual Report to Congress 1979*, **3**, Washington DC, 1980, p. 116.
22. Horwich and Weimer, op. cit.
23. See, for example, James W. McKie, 'Federal Energy Regulation', in *Annual Review of Energy* (eds), Jack M. Hollander and Harvey Brooks, **9**, 1984, Palo Alto, California, Annual Reviews Inc., pp. 321–49.

24. Petroleum Association of Japan, *Petroleum Industry in Japan, 1983*, Tokyo, August 1984, pp. 23–5.
25. Testimony of Donald T. Regan, Secretary of the Treasury in *International Financial Markets and Related Matters*, Hearing before the Committee on Banking, Finance and Urban Affairs, House of Representatives, 97th Congress, 2nd Session, 21 December 1982, Washington DC, p. 64.
26. C. C. Garvin, 'Hard Choices Ahead', *Esso European Review*, London, Spring 1979, p. 1.
27. Petro-Canada, *Canada's National Energy Corporation*, Calgary, Alberta, July 1984, p. 2; and US General Accounting Office, *Petro-Canada: The National Oil Company as a Tool of Canadian Energy Policy*, Washington DC, 15 October 1981.
28. See, for example, Salomon Brothers Inc., *Canada's New Energy Policy*, New York, 1 August 1985.
29. W. H. Hopper, Chairman of the Board, Petro-Canada, *1984 Annual Report*, 31 March 1985, Calgary, Alberta, p. 2.
30. Taro Ishibashi, 'Japan's Oil Exploration and Development Programmes, Financial and Technical Assistance', *United Nations Symposium on Financing of Petroleum Exploration and Development in Developing Countries*, Athens, Greece, April 1985.
31. JNOC, *Japan National Oil Corporation*, Tokyo, 1985, pp. 2–3.
32. *Japan National Oil Corporation*, op. cit., p. 19.
33. See Irvine H. Kuniholm, *Aramco, the United States and Saudi Arabia: A Study of the Dynamics of Foreign Oil Policy, 1933–1950*, Princeton, NJ, Princeton University Press, 1981.
34. See James E. Akins, 'OPEC Action: Consumer Reactions 1970–2000', at the OPEC Seminar on *OPEC and Future Energy Markets*, Vienna, October 1979, p. 7.
35. *Domestic Crude Oil Decontrol*, Washington DC, 1979, p. 414.
36. 'Debate on Government Role Splits Industry: Some US Oil Firms Fear Loss of Controls', *IHT*, 29 December 1980, pp. 7–9.
37. See, for example, Stephen C. Stamos, 'Energy Future Revisited—Adjustment or Restructuring?', *Challenge*, July–August 1984, pp. 55–64.
38. Lord Kearton, 'The Oil Industry—Some Personal Recollections and Opinions', in *The Changing Structure of the World Oil Industry* (ed.), David Hawdon, London, Croom Helm, 1985, p. 11.
39. See, for example, 'Privatisation in Britain', *The Economist*, London, 23 February 1985, pp. 72–4.
40. BNOC's *Report and Accounts for 1984*, 29 May 1984, pp. 8 and 18.
41. See, for example, Michael G. Webb, 'Energy Policy and the Privatisation of the UK Energy Industries', *Energy Policy*, London, February 1985, p. 33.
42. See Albin Chalandon, 'L'État actionnaire n'a jamais su ce qu'il voulait', *L'Expansion*, Paris, 1–11 July 1983, pp. 75–81.
43. Testimony before Subcommittee on Energy and Power of the Committee on Interstate and Foreign Commerce, House of Representatives, 96th Congress, 1st Session, *Domestic Crude Oil Decontrol*, Washington DC, 1979, p. 362.
44. See Alexander G. Kemp and David Rose, 'Comparative Petroleum Taxation', *Petroleum Economist*, February 1983, p. 55.
45. Report on Matters Concerning the Natural Gas Industry in British Columbia, British Columbia Energy Commission, 1973, Part 4, p. 7. Quoted by Kamal Hossain, Director of the Centre for Research on the New International Economic Order in Oxford, *Law and Policy in Petroleum Development*, London, Frances Pinter, 1979, p. 196.
46. Ibid.
47. Report no. 60, 1974–75, Finance Committee, Norwegian Storting (English translation mimeographed), p. 4, quoted by Hossain, op. cit., pp. 192–3.
48. Official Report, Parliamentary Debates, House of Commons, 1974–75, **882**, p. 462—Mr Edmund Dell, Paymaster-General, quoted by Hossain, op. cit., pp. 192–3.

49. Rawleigh Warner, Chairman of Mobil Corp, in 'Mobil Takes on the Media', *International Management*, Maidenhead, England, August 1981, p. 13.
50. Income divided by the average of the beginning and end-of-year direct investment positions.
51. US Department of Commerce, *Survey of Current Business*, Washington DC: 'US Direct Investment Abroad in 1981', August 1982, p. 17; and 'US Direct Investment Abroad in 1983', August 1984, p. 24.
52. 'Esso REP pénalisée pour ses succès', *Esso Informations*, Paris, December 1984, p. 3.
53. Raymond Vernon, 'The Influence of the US Government upon Multinational Enterprises: The Case of Oil', in *The New Petroleum Order*, ed., Antoine Ayoub, Quebec, Les Presses de l'Université Laval, 1976, p. 59.
54. US General Accounting Office, *The Changing Structure of the International Oil Market*, Washington DC, 11 August 1982, p. 49.
55. Tayeh Abdul Karim, in *The Baghdad Observer*, 3 September 1981, reproduced in *Arab Oil and Gas*, Paris, 16 September 1981, p. 31.
56. The Trade Act of 1974 (Section 502, b, 2) and the Trade Agreements Act of 1979 (Section 2462, b, 2).
57. See J. Kaikati and W. A. Label, 'The Foreign Antibribery Law: Friend or Foe?', *Columbia Journal of World Business*, Spring 1980, New York, p. 50.
58. See Federal Reserve Bank of Chicago, *International Letter*, Chicago, 11 April 1980, p. 1.
59. E. J. Frydl, 'The Debate over Regulating the Eurocurrency Markets', Federal Reserve Bank of New York, *Quarterly Review*, Winter 1979–80, p. 16.
60. 'Syndicated Bank Loans', *IHT*, Paris, 3 December 1979, p. 9.
61. 'Political Deadlock with US Continues as Economic War Spreads to London', *Middle East Economic Survey* (*MEES*), 10 December 1979, p. 8.
62. 'Why did Chase move so fast?', *Euromoney*, London, January 1980, p. 11.
63. *MEES*, 24 December 1979, p. 5.
64. Ibid., 10 December 1979, pp. 2–3.
65. See 'Saudi Finance Minister, Sheikh Mohammed Ab-Alkhail on the Fund', *An-Nahar Arab Report and Memo*, Paris, 28 January 1980, p. 8.
66. 'Portions of Official Iranian Assets Previously Frozen in US Financial Institutions', *International Letter*, Federal Reserve Bank of Chicago, 30 January 1981, p. 4.
67. 'Sanctions said to be effective', *Financial Times*, London, 27 May 1982, p. 4.
68. See Jenny Ireland, 'Still viewed by the world as a safe haven', *Financial Times*, 27 September 1982, p. x.
69. See 'Le Caire a "gelé" deux milliards de pétro-dollars arabes', *Le Monde*, Paris, 24 January 1980, p. 5; and 'Interest will be paid on Arab funds', *MEED*, London, 8 February 1980, p. 23.
70. See 'Americans leaving Libya say home is where Qadhafi is—many resent Reagan request that they return to US: "Safer than Philadelphia"', *Wall Street Journal*, New York, 16 December 1981, pp. 1 and 21; and Michael Petrie-Ritchie, 'US/Libya—trade ruling makes life difficult for US firms', *MEED*, December 1982, pp. 26–9.
71. Richard Gilmore, 'Grain in the Bank', *Foreign Policy*, New York, Spring 1980, p. 168.
72. John Edwards, 'Carter's wheat embargo backfires', *Financial Times*, London, 11 March 1980, p. 16.
73. 'The Embargo on US Grain and Phosphate Shipments to the USSR', *International Letter*, Federal Reserve Bank of Chicago, 8 May 1981, p. 2.
74. See, for example, 'Grain-glutted America chews the cud with Russia', *The Economist*, London, 29 May 1982, pp. 69–70.
75. See, for example, 'Pipelines asunder', *The Economist*, 4 September 1982, pp. 25–6.
76. Edward R. Luter, Senior Vice President of Dresser Industries in Michael Blumstein, 'Dresser says unit can finish order', *New York Times*, 30 August 1982, p. D2.
77. See 'Les ventes américaines de haute technologie à l'Union Soviétique sont peu importantes', *Le Monde*, Paris, 31 December 1981, p. 6.

78. See, for example, 'Nigeria Cites Its Oil as Policy Weapon', *IHT*, 6 April 1981, p. 4.
79. 'Arab Oil Ministers Reject Libyan Motion for Oil Boycott of United States', *MEES*, 21 June 1982, pp. 1–3.
80. See Z. Mikdashi, *The International Politics of Natural Resources*, Ithaca, Cornell University Press, 1976, pp. 175–6.
81. Federal Reserve Bank of Chicago, *International Letter*, 30 November 1984, p. 1.
82. See *MEED*, 20 February 1981, p. 4.
83. Ibrahim M. Oweiss, *Impediments to Arab Investment in the United States*, The Industrial Bank of Kuwait, May 1985.
84. 'Whom do we hurt?', *Journal of Commerce*, New York, 1 March 1982, p. 6.
85. 'Economic Sanctions against the USSR', *International Letter*, Federal Reserve Bank of Chicago, 19 November 1982, p. 2.
86. 'URSS—Paris s'oppose à toute remise en cause de la licence GE-Alsthom', *Bulletin de l'industrie pétrolière*, Paris, 27 May 1982, pp. 2–3; and Bruno Dethomas, 'Réquisitionnée par le gouvernement, la filiale français du groupe américain Dresser livrera des compresseurs destinés au gazoduc euro-sibérien', *Le Monde*, 25 August 1982, p. 20.
87. See 'Sanctions—and when they are only halfway up . . .', *The Economist*, 22 January 1983, p. 62.
88. See 'Bonn attacks US pipeline move', *Financial Times*, London, 22 June 1982, p. 16; 'Le blé américain pour l'URSS—ce qui est bon pur les États-Unis n'est pas bon pur les Alliés', *Le Monde*, Paris, 1–2 August 1982, pp. 1 and 3; 'Exporting American sanctions', *The Economist*, London, 26 June 1982, p. 20; 'Siberian pipeline—over to the lawyers', *The Economist*, London, 14 August 1982, pp. 59–60; and 'Canadian Law on foreign embargoes set to take effect', *Financial Times*, London, 3 January 1985, p. 4.
89. Peter Rees, Minister of Trade, interview in *Newsweek*, New York, 18 October 1983, p. 68.
90. This is confirmed by a legal adviser to the Carter Administration, Stanley Marcus; see Leslie H. Gelb, 'Pipeline: an impasse with no end in sight', *New York Times*, 31 August 1982, p. D1.
91. Committee on Government Operations, *The Adequacy of the Federal Response to Foreign Investment in the United States*, House Report, 96th Congress, 2nd Session, Washington DC, 1980, p. 13.
92. Edward R. Luter, 'The Tobago Syndrome: The Extraterritorial Application of US Laws and its Impact on International Trade', in *The International Essays for Business Decision-Makers*, **IV**, ed. Mark B. Winchester, The Center for International Business, Houston, Texas, 1979, p. 43.
93. Ahmad Zaki Yamani, 'The Impact of Oil on International Politics', *MEES Supplement*, Nicosia, 5 May 1980, p. 5.
94. See, for example, 'Should the US be planning for another oil shock?', *Business Week*, 4 April 1983, p. 34.
95. *Arab Oil and Gas*, Paris, 1 November 1979, pp. 19–20.
96. *Saudi Gazette*, Jeddah, 22 December 1980, p. 3.
97. See, for example, 'The conflict over countering the cartel', *Business Week*, New York, 30 July 1979, pp. 56–7.
98. See, for example, 'Military Intervention! What For?', 'US Defence Secretary Repeats Threats of Military Intervention in Oil-Exporting Countries'; and 'To Face Threats of Military Intervention, Arabs May Use Oil Weapon', in *Arab Oil and Gas*, 16 July 1979, pp. 3–4, 20 and 21.
99. See, for example, Joseph S. Nye jr., 'Energy Nightmares', *Foreign Policy*, New York, Fall 1980, p. 148.
100. See Robert Bailey, 'Superpower war games muddy Gulf waters', *MEED*, London, 28 March 1980, p. 8.

101. Mana Saeed Al-Otaiba, address at OPEC Seminar, *'OPEC and Future Energy Markets'*, Vienna, October 1979, p. 4.
102. André Giraud, 'De la confrontation à la co-opération pour résoudre les problèmes de l'énergie', *Symposium de Davos*, European Management Forum, Switzerland, 7 February 1980.
103. Henry Kissinger, *Years of Upheaval*, London, Weidenfeld and Nicolson, Michael Joseph, 1982, p. 921.
104. J. W. Hopkins, 'Recent experience with the IEA's International Sharing System', *The OECD Observer*, Paris, May 1981, pp. 8–10.
105. The Comptroller General, Report to the Congress, *The United States Remains Unprepared for Oil Import Disruptions*, **2**, US General Accounting Office, Washington DC, 29 September 1981, pp. vii–31.
106. David Howell in *The Guardian*, London, 3 January 1980, p. 1.
107. EC, *The European Community and the Problem of Energy*, Luxemburg, 1983; and George Blondel, Director for Oil and Gas at the European Commission, in 'Sharing the burden of Middle East oil imports', *MEED*, 13 July 1985, pp. 32–8.
108. Blondel, op. cit., p. 34.

5 Financial Aspects

This chapter will review some of the complex relationships which exist between oil and the financial markets. In particular, we will look at the effects that the post-1973 oil events had on financial markets, and the principal protagonists of the oil business—oil-importing countries, oil-exporting countries, and transnational enterprises. Close attention will be devoted to major aspects of financing oil.

SIGNIFICANCE OF OIL TO FINANCIAL FLOWS

Oil has been an important factor in influencing international financial flows, especially since the price surges of the 1970s. These oil price changes produced one of the most massive and sudden transfers of income and wealth among nations in modern history. One may have to go back to post-First World War reparations to find transnational transfers of resources on a comparable scale. Another historical parallel was the massive transfer of bullion to Spain following the conquest of the Aztec and Inca empires in the sixteenth century.

Today (1985) some 25 million barrels of oil are exported per day, and this represents some 20 per cent of total world trade in goods. Any dollar change in the price of oil thus has a direct impact on international financial flows, now put at over $US9 billion per annum. Since world oil production is over twice the size of internationally traded oil (i.e. 55 million b/d), the direct transfer impact (between producers and consumers) of a change of $1 in the oil price could reach some $20 billion per annum. Unless offsets occur, the cumulative spin-off impact could be much greater than such a direct change. The indirect repercussions would affect other sources of energy and oil-dependent or oil-related industries. These include the construction of drilling equipment, of transport means (pipelines, tankers, etc.), storage, refining and chemical plants, metallurgical and automotive industries, contracting, and other sectors.

One can illustrate the direct impact of an oil price change by reference to the world's largest oil-consuming country, namely the United States. Estimates in Table 18 indicate a massive transfer of $124 billion in income to oil producers, national and foreign, following the $20-a-barrel price increase of 1979–80. The transfer of income in favour of foreign producers then

Table 18 Impact of oil prices on US economy

	1974–75	1979–80	1983	1984[e]
1. Change in oil price, per barrel	+$8	+$20	−$5	−$2
2. US oil product consumption, billion barrels per year	6.1	6.2	5.5	5.9
3. Aggregate US effect = (1) × (2), billions	$49	$124	−$28	−$12
4. (3) as percentage of GNP	3.7%	4.7%	−0.9%	−0.3%
5. US oil imports, billion barrels per year	2.3	2.5	1.8	2.0
6. Aggregate import effect = (1) × (5), billions	$18	$49	−$9	−$4
7. (6) as percentage of GNP	1.4%	1.9%	−0.3%	−0.1%

[e] Estimate.
Source: Department of Energy, Bureau of Economic Analysis, quoted by Goldman Sachs, *The Pocket Chartroom*, New York, November 1984, p. 1g.

reached an estimated $49 billion. The impact on inflation was given at about 1.9 per cent.

More difficult to measure, but no less significant, are the indirect effects of oil price increases, including the policies of adjustment or austerity which culminated in the severe recession of 1982–83. The acuteness of the latter, varied, however, depending, *inter alia*, on the impact of fiscal and monetary policies of individual oil-importing countries. Empirical analyses have attempted to identify the nature of the link between oil price peaks and economic recessions. Results indicate an association, though a causal connection is not fully proved.[1]

Given the magnitude and the precipitate nature of such a global financial transfer in the 1970s and early 1980s in favour of a few oil-exporting countries, various parties in the oil-importing countries had deep fears; in particular that:

— the growth of oil-importing economies could be choked by onerous oil import bills;
— the international financial system could not cope with the financial imbalances created by the oil price increases;
— surplus capital oil-exporting countries could use their newly acquired wealth to control leading Western enterprises or try to disrupt exchange markets by sudden transfers of funds from one currency to another.

The oil trade has also affected exchange rates. As an illustration, one could refer to the strength of sterling in the early 1980s, due in no small measure to rising oil prices and exports. In 1983 the Exchequer was collecting about £8 billion annually, and the oil sector was contributing some 5 per cent to the United Kingdom's GNP and £2.3 billion to the United Kingdom's balance of payments current account.[2]

While oil remains an important factor in trade flows—having a significant impact on the balance of payments, exchange rates and other indicators of economic performance—capital flows are thought (in the mid-1980s) to be ten to fifteen times the size of trade flows. These financial flows (mostly influenced by real-yield differentials of financial and tangible assets placed in different convertible currencies, and future expectations of returns) have played a crucial role in influencing exchange-rate levels.[3] Nevertheless, there are other forces at work, besides yield, which are difficult to measure. Chief among these is the 'refuge' value of a currency in terms of economic and political security.

The following section examines the impact of oil price changes on selected oil-exporter countries. It focuses in particular on those countries which managed to accumulate financial assets by the 1980s; namely, the few sparsely populated Arab Gulf countries. It further reviews their investment policies.

OIL EXPORTERS' FINANCIAL STRATEGIES

In the 1970s, Western countries had worries about potential political or economic pressures which oil-exporting countries could exercise by means of their foreign investments. These worries could partly be explained by certain actions of oil-exporting countries; for example:

- the embargo imposed on selected destinations by Arab oil-exporting countries in the wake of the October 1973 Arab–Israeli hostilities;
- Iran's threats during the US embassy hostage crisis to upset the dollar exchange market by transferring its dollar assets into other currencies;
- more generally, much rhetoric about the use of oil as a political weapon, especially in the 1970s.

In the deployment of their accumulated pools of external funds, oil exporters have generally given high priority to safe outlets in Western financial markets (Table 19). A small number of countries account for the major portion of the external financial assets of oil-exporting countries accumulated over the 1970s and 1980s. In the mid-1980s the largest net accumulations belonged to Kuwait, Saudi Arabia and the United Arab Emirates.

Lacking the requisite financial institutions, appropriate talents and familiarity with debtor countries, in addition to their being generally averse to assuming lending risks, these countries initially channelled most of their surplus funds into transnational commercial banks, into government securities of industrial countries and, to a lesser extent, into intergovernmental financial institutions such as the World Bank and the International Monetary Fund. A large portion of oil-exporting countries'

Table 19 Deployment of oil exporters' surplus, 1974–1983 ($m.)

	Flows										Levels[a]		
	1974	1975	1976	1977	1978	1979	1980	1981	1982	1983	1973	1979	1983
ANALYSIS BY COUNTRY													
United Kingdom:													
Sterling bank deposits	1.7	0.2	−1.6	0.3	0.1	1.5	1.4	0.5	1.2	0.1	1.3	4.2	5.4
Eurocurrency deposits	13.8	4.1	5.6	3.1	−6.7	15.5	14.5	7.8	−9.2	−7.0	5.3	46.2	50.3
Government paper	3.6	−0.5	−0.9	—	−0.4	0.2	2.0	0.8	0.1	0.2	0.7	2.1	3.5
Direct investment	0.2	—	—	—	0.1	—	—	—	0.1	—	—	0.3	0.5
Other investments[b]	1.7	0.5	1.3	0.6	—	0.2	0.2	0.2	−0.7	−0.5	—	4.7	3.4
Total	21.0	4.3	4.4	4.0	−6.9	17.4	18.1	9.3	−8.5	−7.2	7.3	57.5	63.1
United States:													
Bank deposits	4.2	0.6	1.9	0.4	0.8	5.1	−1.3	−2.0	4.6	0.9	2.4	15.5	17.7
Treasury bonds and notes	0.2	2.0	4.2	4.3	−1.5	−1.2	8.2	10.9	6.9	−5.5	0.1	7.5	28.0
Treasury bills	5.3	0.4	−1.0	−0.9	−0.9	3.3	1.4	−0.4	0.4	−1.2	0.4	6.6	6.8
Direct investment	0.1	—	—	—	0.1	—	0.2	2.9	0.5	—	0.1	0.4	4.0
Other investments[b]	1.7	4.9	6.0	3.5	1.9	−0.2	8.6	6.4	0.3	−3.7	0.5	17.8	29.4
Total	11.5	7.9	11.1	7.3	0.4	7.0	17.1	17.8	12.7	−9.5	3.5	47.8	85.9
Germany:[c]													
Bank deposits	0.7	0.4	0.5	0.8	−0.2	0.6	1.2	0.1	−1.6	−0.4		3.9	2.2
Government paper	0.1	0.6	0.4	0.3	−0.4	—	6.8	6.3	1.4	−0.3		1.5	11.5
Direct investment	0.5	0.1	—	—	0.3	0.1	0.1	—	—	—		0.5	0.3
Other investments[b]	0.2	0.9	0.4	0.9	0.8	1.0	1.4	1.2	−0.6	−0.7		9.7	7.6
Total	1.5	2.0	1.3	2.0	0.5	1.7	9.5	7.6	−0.8	−1.4		15.6	21.6

Other industrial countries:[d]													
Bank deposits	8.2	4.6	5.6	13.3	13.3	17.9	28.5	−3.9	−7.3	−4.9	—	61.5	64.9
Other investments[b]	2.6	5.8	5.1	4.5	3.0	2.5	7.9	12.2	5.9	−0.6	0.6	18.6	52.4
IMF and IBRD[e]	2.4	5.0	1.7	0.4	−0.3	−0.8	1.7	2.1	2.1	4.3	2.2	10.5	20.8
OEC bank credit to non-banks worldwide[f]	—	1.5	1.6	1.5	1.6	0.1	1.7	2.1	0.1	2.8	—	6.0	13.0
Placements with LDCs	4.9	6.5	6.4	7.0	6.2	8.7	6.3	7.2	3.9	1.2	—	39.7	58.3
Total identified net cash surplus	52.1	37.6	37.2	40.0	17.8	54.5	90.8	54.4	8.1	−15.3	13.6	257.2	380.0
PORTFOLIO STRUCTURE													
Bank deposits	28.6	9.9	12.0	17.9	7.3	40.6	44.3	2.5	−12.3	−11.3	9.0	131.3	140.5
Government and other public sector paper:													
UK Treasury bills	2.7	−0.9	−1.1	−0.3	0.1	—	−0.1	−0.1	—	—	0.7	0.4	0.1
US Treasury bills	5.3	0.4	−1.0	−0.9	−0.9	3.3	1.4	−0.4	0.4	−1.2	0.4	6.6	6.8
US Treasury bonds and notes	0.2	2.0	4.2	4.3	−1.5	−1.2	8.2	10.9	6.9	−5.5	0.1	7.5	28.0
British government stock	0.9	0.4	0.2	0.3	−0.5	0.2	2.1	0.9	0.1	0.2	—	1.7	3.4
German public sector[g]	0.1	0.6	0.4	0.3	−0.4	—	6.8	6.3	1.4	−0.3		1.5	11.5
Total	9.1	2.5	2.7	3.7	−3.2	2.3	18.4	17.6	8.8	−6.8	1.2	17.7	49.8
Non-foreign exchange reserves:													
Gold	0.1	−0.1	0.1	—	0.2	—	0.1	—	−0.1	—	1.4	1.7	1.7
SDRs	—	—	—	0.1	0.2	0.8	0.2	0.5	0.3	−0.6	0.4	1.4	1.7
Reserve position in IMF	1.9	2.8	1.2	0.2	−0.7	−1.8	1.3	1.5	0.8	4.4	0.4	4.0	12.0
IBRD bonds	0.4	2.3	0.4	0.1	—	0.2	0.1	0.1	1.1	0.5	—	3.4	5.4
Total	2.4	5.0	1.7	0.4	−0.3	−0.8	1.7	2.1	2.1	4.3	2.2	10.5	20.8
Equities, property and other investments[h]	7.0	13.7	14.4	11.0	7.8	3.7	20.1	25.0	5.6	−2.7	1.2	58.0	110.6
Placements with LDCs	4.9	6.5	6.4	7.0	6.2	8.7	6.3	7.2	3.9	1.2	—	39.7	58.3
Total	52.1	37.6	37.2	40.0	17.8	54.5	90.8	54.4	8.1	−15.3	13.6	257.2	380.0

[For Notes and Source see p. 000]

current account surpluses was placed in short-term papers and bank deposits. The rise and fall of this category of assets reflected the performance of these countries' current accounts. To illustrate, one could attribute the heavy disinvestment in this category of assets in 1983 (see Table 19) to the need for funding large current account deficits, notably those of Saudi Arabia in that year (Table 20). One underlying assumption of Saudi official investment behaviour has been that the current account surpluses are temporary. Oil exporters' current account deficits are largely due to declining oil revenues (resulting from a fall in oil export volumes and

Table 20 OPEC current account balances*

($ billion)	1984	1983	1982	1981
Saudi Arabia	−12.0	−18.4	−1.1	38.4
Kuwait	7.0	4.6	4.5	13.7
Iran	—	0.1	2.6	−4.0
Iraq	−4.0	−3.4	−11.6	−8.4
UAE	3.0	3.5	5.7	10.7
Nigeria	−1.0	−4.7	−7.7	−5.9
Libya	—	−1.7	−1.6	−4.3
Algeria	—	−0.1	−0.2	0.1
Venezuela	4.0	3.7	−4.2	4.0
Indonesia	−4.0	−6.3	−5.3	−0.6
Qatar	—	2.0	2.5	4.3
Gabon	0.4	0.1	0.3	0.4
Ecuador	−0.2	−0.1	−1.2	−1.0
Total OPEC	−6.8	−20.7	−17.3	47.4

* Based on IMF data and Shell estimates.

Source: Shell Briefing Service, *Oil, Gas and Coal in 1984*, 2, London, 1985, p. 6.

Notes and source to Table 19

— Not available.

[a] The year-end levels are only partially derived from stock data. Where this information is not available the levels have been estimated by accumulating flows.

[b] Mainly loans and holdings of equities.

[c] Figures for Germany relate only to the members of OPEC and do not include deployment by Bahrain, Brunei, Oman, and Trinidad and Tobago—the other oil exporting countries (OEC).

[d] Includes Austria, Belgium, Canada, Denmark, Finland, France, Ireland, Italy, Japan, Luxemburg, Netherlands, Norway, Spain, Sweden and Switzerland.

[e] Includes holdings of gold valued at SDR 35 per fine ounce.

[f] Derived from IFS data but there may be a small element of overlapping with data included under other headings.

[g] Schuldscheine unquoted medium-term promissory notes issued by the federal government, local authorities, etc.

[h] Includes OECs' bank credit to non-banks worldwide derived from IFS data (see footnote [f] above) and some unidentified government paper.

Source: Bank of England, *Quarterly Bulletin*, London, March 1985, p. 71.

in prices) coupled with comparatively large imports (Table 21). Heavy military expenditure, mismanagement and unwise commitments to domestic prestige projects have significantly contributed to the deepening of oil exporters' deficits and to depleting their external financial assets.

Table 21 Estimated OPEC oil exports and revenues

	Exports (million b/d)	Revenues ($ billion)
1984	15	159
1983	15	160
1982	17	202
1981	21	253
1980	25	279
1979	29	197
1978	28	133

Source: Shell Briefing Service, *Oil, Gas and Coal in 1984*, 2, London 1985, p. 6.

Available evidence suggests that oil exporters have resorted to increased diversification of their assets among markets. This trend may in part reflect the gradual liberalisation of markets outside the United States and may in part reflect a desire to avoid the risks of concentrating assets in a single currency, domicile or category of assets. In the 1970s, investors from oil-exporting countries had favoured very small levels of participation in foreign companies (less than 5 per cent) for diverse reasons. Chief among these was their limited knowledge of foreign markets, their desire to have a portfolio of diversified investments which could be readily disposed of, and the wish to avoid transparency required by law for larger holdings. The wish for discretion was prompted by a concern to avert possible adverse reactions from the public, partisan interests or inquisitive government bodies in capital-importing countries.

In the 1980s growing confidence and expertise among investors from oil-exporting countries, and the reduced external hostility to their investments in host countries, have made them more forthright in taking larger shares in, or even straight control of, foreign enterprises. This is notable of Kuwaiti investors. Moreover, the national or regional financial institutions of oil-exporting countries have developed skills and confidence over the years to enable them to participate directly on a more significant scale in international lending and other transnational financial transactions.[4]

Western financial markets have generally not been disturbed by oil exporters' investment policies. Investors from the government sector of oil-exporting countries have not attempted to influence the actions of corporations or banks to suit their special foreign policy objectives. Even if one were to assume that a given investor wanted to 'punish' a specific bank by transferring his deposits, that bank could obtain refinancing from the market. In the unlikely event that the depositor would ask for his funds in cash, the impact on the world financial markets is bound to be negligible. Indeed, oil exporters' investments in Western markets are relatively small compared to the size of the markets concerned, and to other foreign investments from the industrially developed countries. Moreover, institutions and markets have a whole set of instruments which could deal effectively with any attempt by foreign investors to upset their operations.[5]

FUNDING OIL IMPORTS

In the 1970s and 1980s oil was a key factor in external indebtedness. Oil-related indebtedness in that period was due to (a) higher import bills, especially after the price increases of the 1970s, and (b) borrowing to finance oil and other energy projects, in the hope of cutting down the oil import bill or producing oil for export within a few years. With reference to oil import bills, non-oil developing countries' external debts rose by about $482 billion from 1973 to 1982. One analyst, William R. Cline,[6] attributed most of the increase (i.e. $401 billion) to external causes. The major components were estimated as follows:

— $260 billion, directly traceable to oil price increases;
— $100 billion, attributable to a deterioration in the terms of trade (i.e. a decline in the price index of exports relative to that of imports), as well as to reduced export volume;
— $41 billion, ascribable to high real interest rates.

Other analysts (e.g. the IMF economist Eduardo Wiesmer, a former Colombian finance minister) attributed the expansion of developing countries' debt, especially in Latin America, largely to internal economic mismanagement, particularly, excessive fiscal deficits.[7]

Indeed, these debtor countries maintained their levels of public expenditure, including support for artificially low prices for certain commodities and low interests on certain types of credit—thanks to large subsidies. Moreover, despite increases in import prices, these countries also sought to maintain their programmes of socio-economic development, instead of adjusting to the adverse external economic conditions. Finally, economic mismanagement covered areas from 'white elephants' to misappropriation. All these factors contributed to ballooning fiscal deficits in the 1970s.

Foreign debt need not have increased so dramatically between 1973 and 1982, despite the oil price shocks, if appropriate action had been taken by oil-importing developing countries, by financial intermediaries and by industrial countries. While not minimising the adverse impact of the oil price shocks (especially those of 1979–80) on the balance of payments of a number of oil-importing countries, improvident, unrealistic and wasteful government policies have also been responsible for excessive indebtedness. Several middle-income, industrialising, developing countries thus expanded their commercial debt commitment in order to maintain growth rates, consumption and import levels—including oil imports, despite price increases. Debtors were further encouraged by three factors:

— the ready availability of loanable funds from commercial sources, which exempted these countries from the usual 'austerity programmes', as required by the International Monetary Fund (IMF);
— the moderate or even negative real interest rates prevailing in the 1970s;
— expectations of impending recovery in the world economy (which proved unrealistic, as the recession continued into the later 1970s and the first half of the 1980s).

The application of timely and genuine domestic adjustment policies—espousing realistic market-related prices, exchange rates and interest rates, as well as balanced budgets, moderate monetary expansion, appropriate incentives for productive investments, and avoidance of waste—would have alleviated the pressure of domestic demand, including the demand for oil. As the Bank for International Settlements (BIS) candidly stated:

> Latin American countries whose residents have been exporting capital on a massive scale would be well advised to put their house in order so as to end this capital outflow and even, hopefully, to draw some of the flight capital back. It is too much to expect the rest of the world—whether multilateral institutions, governments, banks or corporations—to perform the development functions that these countries' own nationals refuse to assume.[8]

A Federal Reserve Bank source estimated the flight of capital from five major Latin American debtor countries (Argentina, Brazil, Chile, Mexico and Venezuela) at over $84 billion between 1974 and 1982, while the BIS estimated the flight of capital from Latin America at some $50 billion between 1978 and 1982.[9]

Commenting on the developing countries' foreign debts, an international banker stated that 'in 9 cases out of 10 . . . the problem is created by something the government is doing domestically, which leads to a weak balance of payments, regardless of the international situation'.[10] A key domestic policy error is usually to combine high fiscal deficits with an overvalued exchange rate. When the deficit is ultimately monetised, inflation rises and a vicious spiral starts. Indeed, through appropriate domestic policies, a number of countries heavily dependent on oil imports for their energy requirements—whether developed (e.g. Japan) or developing (e.g. South Korea)—managed to ride the oil price increases without undue burdens on their balance of payments.

Financial institutions were equally responsible during the 1970s for encouraging selected developing countries to incur excessive debts, some with inappropriate features such as general-purpose balance-of-payments financing, with maturities crowded within short periods. With reference to the former criticism, it has to be noted that capital is fungible, and as such it is difficult for commercial banks to ascertain that specific project lending is not used in areas other than those it was intended for. Other inter-governmental institutions, however, such as the World Bank, can exercise a measure of effective control in ensuring that credit facilities are effectively used in the agreed projects.

The race in the 1970s to lend to the advanced developing countries was due to the banks' wish to compensate for slack credit demand in developed markets, and their need to recycle the sizeable OPEC deposits.

Industrial countries' policies have also had a share in the responsibility for the indebtedness problem of developing countries. Three points deserve to be raised on this subject:

- Financial institutions, on the acquiescence or even the overt encouragement of their governments, had in the 1970s and early 1980s recycled oil exporters' surplus funds to oil-importing countries.
- In a world of increasingly integrated financial markets, the domestic policies of 'reserve currency' countries affect other countries. This is notably the case with the United States, whose massive budget deficits between 1980 and 1985 (coupled with a non-expansionist monetary policy) have been instrumental in hiking real interest rates on dollar assets, with a peak in 1979–80. Interest rate increases were borne by major debtor countries whose commercial debt was largely contracted at variable interest rates (i.e. adjustable periodically in relation to certain prevailing market rates indices). Despite a decline in the nominal US dollar price of oil between 1982 and 1984, the considerable increase in the value of the US dollar in terms of other currencies had on balance increased the price of oil outside the United States (see Figure 2). This has aggravated the financial position of debtor oil-importing countries.
- Interest on foreign debt can only be serviced, and the principal amortised, out of export proceeds. If debtor countries are denied market outlets by industrial countries which also happen to be the major creditor countries, debtors are in a bind. Commenting on the developing countries' external debt problem, the Bank for International Settlement acknowledged that 'Barriers aimed at imports of industrial and agricultural products from the LDCs are more than a grave danger—they are an explosive one. . . . A lasting solution is inconceivable without a vigorous development of LDC exports to the industrial world'.[11]

FUNDING OIL OPERATIONS

Risks and Corporate Financing

The time allowed for the completion of energy projects (often several years) is based on various assumptions about future energy consumption and supply patterns, energy prices, profit margins, competition and government policy (taxation, depletion or production rates, equity participation, labour laws, environmental and other regulations). Risks involved determine the size and nature of funds needed. These cover five main areas:

— Geological risks. Costly exploration could end up in dry holes or in oil or gas which could prove uneconomical to produce.
— Commercial risks. Unpredictable fluctuations in supply, demand and/or in prices could have devastating economic effects on energy projects; thus a drastic fall in oil prices could destroy the commercial viability of higher-cost energy projects (often capital intensive, calling for substantial investment over long gestation periods reaching up to fifteen years and occasionally beyond).
— Monetary and foreign exchange risks; that is, uncertainty regarding exchange-rate fluctuations, the imposition of exchange controls and restrictions on the repatriation of capital or profits.
— Political risks; that is, adverse changes in government policy or legislation, social unrest or armed conflict, damage to property rights, contract frustration and other confiscatory or arbitrary measures.
— The absence of an adequate legal framework which would lead to excessive delays in negotiations and permit administrative interference, stifling bureaucracy and rampant corruption.

Thus, the financing of most energy projects normally depends on stable funds which accept a relatively high level of risk-taking. Major components of such funds would be: fresh equity capital; internal financing, derived from profits, depreciation and/or amortisation of capital expenditure; sale of assets; government funding or government-guaranteed credits. The financial challenge of oil companies has been to organise an optimal mix between debt and equity. What ratio is optimal depends on the size and stability of a company's prospective cash flow. With the ever-looming risk of dry holes, oil companies should be wary of resorting to excessive debt-financing, lest falling income or losses prevent the company from honouring interest payments and the repayment of the principal. For an indication of the international oil industry's reliance on financial markets, one can refer to borrowing by major oil companies. Table 22 shows that borrowing by twenty-four major oil companies increased in nominal terms by a factor of 2.75 over twelve years (1971–82). Yet the 'debt to total-capital-employed' ratio, for the same companies and for the same period, hovered at around 22.5 per cent. A considerable jump in this ratio (i.e. by 7.3 per cent)

Table 22 Major companies' assets, return and debt financing, 1971–1984*

	Total assets $ billion	Net income/ equity† (%)	LT debt $ billion	LT debt/ capital employed‡ (%)
1971	121.7	10.9	19.8	21.8
1972	128.7	9.7	21.1	22.4
1973	146.4	15.5	22.0	21.4
1974	188.3	19.2	24.6	20.9
1975	196.1	12.4	29.7	23.9
1976	215.8	13.5	35.1	26.1
1977	241.8	13.8	38.8	25.8
1978	267.5	13.1	39.4	24.4
1979	330.8	24.0	42.0	22.0
1980	388.1	22.3	45.6	20.5
1981	411.8	16.7	50.6	21.3
1982	423.5	12.1	54.4	21.8
1983	360.2	12.9	43.4	20.5
1984	380.4	12.3	61.4	27.8

* Twenty-four companies over 1971–82, and nineteen companies in 1983–84.
† Includes preferred stock, common stock and surplus.
‡ Includes long-term debt, preferred stock, common stock, surplus and equity of minority interests.
Sources: Chase, *Financial Analysis of a Group of Petroleum Companies, 1982*, New York, April 1984, pp. 22–5; and *Financial Analysis of a Group of Petroleum Companies, 1984*, New York, July 1985, pp. 24 and 26.

occurred over 1983–84 for nineteen major companies, largely due to finance acquisitions or repurchases of their own stocks.

Companies which place excessive reliance on debt financing find themselves in a very dangerous situation once cash earnings dry up. This is dramatically illustrated by the fate of the Canadian multi-billion Dome Petroleum Limited. By mid-1983, Dome's shareholders' equity of C$321.4 million was totally wiped out by losses, while its total assets of C$8,083 million were financed exclusively by debt. At the end of the first quarter of 1985, its shareholders' deficit was C$415 million.[12]

Dome borrowed from banks to finance its various exploration activities (including the expensive and unproductive search for oil in the Arctic waters of the Beaufort Sea); it also borrowed to acquire another company, Hudson's Bay Oil and Gas. Financing that acquisition through debt drove an already highly leveraged enterprise into default.[13]

A distinction can be made here between the financing of exploration, and that of development. The former, owing to its uncertain outcome, is customarily financed out of fresh equity for a new venture, or out of internally generated funds for a going concern. Development, on the other hand, lends itself to some degree of debt financing once commercial reserves are reasonably established, and once development costs are adequately estimated to be at competitive levels.

Financing Acquisitions

The spate of acquisitions and mergers of the 1980s has also drawn substantial funds from financial markets. Table 23 illustrates the multi-billion-dollar acquisitions/mergers in the US oil sector.

Table 23 Major oil acquisitions in the United States, 1979–1985

Acquiring/acquired firms	Terms	Year	Cost $ billion
Chevron/Gulf	100% cash	1984	13.3
Texaco/Getty	100% cash	1984	10.1
Dupont/Conoco	40% cash 60% stock	1981	7.4
US Steel/Marathon	51% cash 49% notes	1981	6.4
Mobil/Superior	100% cash/debentures	1984	5.7
Occidental/Cities Services	45% cash 55% notes	1982	4.0
Shell Oil/Belridge	100% cash	1979	3.6
Royal Dutch-Shell/Shell Oil's minority interests	100% cash	1984	5.7
Fluor Corp./St. Joe Minerals	45% cash 55% stock	1981	2.6
Philips/General American Oil Co.	100% cash	1983	1.1
Philips/Reynold's Aminoil of Texas	100% cash	1984	1.6
Diamond Shamrock/Natomas	common & preferred stock	1985	1.5
Allied Managers/Union Texas Petroleum (50% interest)	65% cash 35% preferred stock	1985	1.4

Sources: Company reports; US Senate Committee on the Judiciary, *Oil Merger Activity*, Washington DC, 15 March 1984, pp. 46–7, and various trade sources such as: D. O. Croll, 'Merger wave hits majors', *Petroleum Economist*, London, April 1984, pp. 125–7; *Business Week*, New York, 13 May 1985, p. 17; and Bernard J. Picchi, *The Structure of the US Oil Industry*, New York, Salomon Brothers Inc., July 1985.

The acquisition/merger strategy was generally prompted by two principal sets of corporate motivations: (a) to avoid predatory take-over attempts, thereby opting for a 'friendly' merger and (b) to secure the advantages of balanced operations, notably through vertically and geographically integrated activities. Company management has sought a friendly merger to avoid predatory take-overs which would aim at the partial or total dismemberment of a company; this is the case when the raider considers the break-up of the company and its liquidation as yielding stockholders a higher value than that obtaining in the stock-exchange market. Company management has also sought mergers, balancing their operations, especially to make up for:

— loss of equity crude-oil sources (for example, in the 1970s after OPEC countries took over producing assets held by concessionaires—mostly the major transnational companies);
— dwindling crude-oil reserves, especially when confronted by rising exploration–development costs.

In the case of the world's biggest oil acquisition—Chevron/Gulf—the Board of Directors of Gulf Corp. recommended a friendly merger with Chevron to their stockholders in 1984 in order to counter the 'corporate raider', T. Boone Pickens, jr. The latter attempted to 'liquidate' Gulf Corp., according to James E. Lee, Chairman of Gulf, notably by asset spin-off. However, Pickens considered that his strategy for bidding for control of various companies was in the interest of stockholders, including himself. He considered his strategy to be beneficial to stockholders since it offered them a price closer to the assessed value of stocks.[14] Indeed, assessed values happened then to be higher than prevailing market values. In the three companies illustrated in Table 24, two vertically integrated companies (Gulf Corp. and Getty) had market values much below liquidation values, (40 and 33 per cent, respectively) as compared with the non-integrated oil and gas producer Superior (84 per cent).

Table 24 Comparative economics of oil mergers

Gulf	Getty	Superior	
$80	$128	$45	Acquisiton price (per share)
$45	$50[e]	$39	Previous share price (minority)*
			Asset/liquidation value†
$113.90	$153.15	$46.20	Total company (per share)‡
$71.40	$120.40	$36.80	Oil, gas and acreage (per share)‡
.40	.33[e]	.84	Market price/liquidation value ratio

[e] Estimated.
* Three months prior to acquisition announcement.
† J. S. Herold Report.
‡ Less liabilities and preferred stock.
Source: Edfred L. Shannan, jr., Chairman and Chief Executive of Santa Fe International, 'Merger Activities in the United States Petroleum Industry', in *Dinar*, The Commercial Bank of Kuwait, 1985, p. 42.

Figures in Tables 24 and 25 are too aggregative, and comparisons based on per-barrel prices can be misleading. Such comparisons overlook the mix of gas and oil in the companies' reserves, the quality or location of these reserves, the production profile of petroleum fields, fiscal levies or incentives, and government regulations. Figures on proved oil and gas reserves are imprecise. They represent estimates of economically recoverable reserves based on current economic, technological and operating conditions. Imprecision is due to incomplete geological information, and variations in estimates depend not only on technology used but also on the timing of production.

Informed observers considered that the economics of producing Superior's oil reserves were more attractive than those of Gulf or Getty. Moreover, the former company was not saddled with low-performing downstream oil assets or other unprofitable investments. Finally, the market may have placed insufficient confidence in the management of Gulf and Getty.[15] It is also possible that reserves are worth more to the buyer than to the seller, for a variety of reasons, such as a technological edge in the recovery of a larger portion of discovered reserves by the buyer, or fiscal-credit advantages.[16]

T. B. Pickens' 'raider strategy' can be illustrated by reference to take-over attempts *vis-à-vis* Philips, Gulf Oil Corp., Cities Service, General American Oil, Unocal, Superior Oil, and others. Although his group neither managed nor wanted to take full control of these companies, his strategy generally yielded a higher value for his groups' stocks by forcing the target companies to find 'friendly' acquirers to buy back their stocks or to restructure by selling off unprofitable assets. The last two options were adopted by various US oil companies. In the case of the Atlantic Richfield Company (ARCO), its board decided to sell poorly performing assets in 1984 (e.g. its metal operations) and to repurchase 10 per cent of Arco's outstanding common stock, 'because we believe Atlantic Richfield's stock to be excellent value at recent prices'.[17]

The three major US companies which concluded oil acquisitions in 1984—Chevron (Gulf), Texaco (Getty) and Mobil (Superior)—increased their oil and gas reserves by 109, 72 and 20 per cent respectively. The price of acquired companies when applied to oil and gas alone (i.e. without reference to other assets) came out for the three acquiring companies at less than 60 per cent of their weighted average of the finding cost of an oil barrel over the 1978–83 period in the United States (Table 25).

To finance their acquisitions, oil companies resorted to heavy borrowing. For example, Mobil's long-term borrowing soared from $5,162 million at the end of 1983 to $11,057 million at the end of 1984, to complete the acquisition of Superior Oil Company at the cost of $5.7 billion. This led to a bulge in Mobil's debt–equity ratio, which rose from 37 to 81 per cent.[18] The rating of Mobil's debentures by Standard and Poor fell from AAA to AA.[19] The debt–equity ratio rose to about 35 per cent in 1984 for four oil companies after their acquisition. These were Chevron, Texaco, Mobil and Philips.[20] The industry debt–equity ratio for major companies (inclusive of the four just mentioned) had hovered, by comparison, at around 22 per cent

Table 25 Major acquisitions reserves

Chevron-Gulf	Texaco-Getty	Mobil-Superior	
2,753	3,261	5,203	Total reserves of acquirer (MMBCOE)*
			Reserves added
2,046	1,881	314	— Oil (MMB)
943	470	708	— Gas (MMBCOE)
2,989	2,351	1,022	Total reserves added
5,742	5,612	6,225	Total post-acquisition reserves (MMBCOE)
109%	72%	20%	Percentage increases in reserves
$4.48	$4.30	$5.58	Price per BBL added
$10.26	$10.72	$10.02	1978–83 average US finding cost (per BOE)*

Sources: Salomon Brothers; Edfred L. Shannon, jun., 'Merger Activities in the United States Petroleum Industry', in *Dinar*, The Commercial Bank of Kuwait, 1985, p. 42.
* MMBCOE = millions of barrels of crude oil equivalent.

over the decade 1972–82 (see Table 21). Having borrowed excessively to buy a company, the buyer is later obliged to increase its retained earnings or to sell assets to pare down debt. However, shedding assets at attractive terms is not a simple process, as was experienced by Chevron.[21]

Bankers have often been called upon to advise on the formulation of corporate strategies in the oil industry and to act as intermediaries; for example, in mergers, acquisitions, divestments, capital funding and restructuring. Thus, Chase Manhattan was the agent bank for a $4.3 billion financial package which enabled Texaco to take over Getty Oil Co. in 1983. Bank of American Corp. wound up in 1984 as the lead bank for Chevron's winning bid for the Gulf Corp. The acquisition/merger strategy has three principal consequences. First, tens of billions of dollars of equity funds have been replaced by debt. Yet equity resources are essential for the risk-taking activities of the oil industry, notably for funding expensive and uncertain programmes. With more debt, an operator is normally bound to take less risk.

Second, in the same way that vulnerability to economic decline or an oil-price downturn reduced the credit standing of these companies, so did their considerably increased indebtedness (notably to pay for their acquisition). This was formally acknowledged by George M. Keller, chairman and chief executive officer of Chevron, which bought Gulf for $13.3 billion in March 1984, by raising its ratio of total-debt to total-debt-plus-equity over 1983–84 from 12 to 48 per cent, thereby lowering the credit standing of Chevron from AAA to AA.[22] To moderate the bulge in their excessive indebtedness as judged by their debt–equity ratio, and to comply with regulatory requirements, some assets of the acquired business would have to be shed.

Third, there is likely to be a tying-up of capital in lines of credit in preparation for take-overs, which may or may not materialise. Such tying-up of credit is bound to tighten the financial markets' liquidity and firm up interest rates—at least for the tie-up period. Viewed with reference to total market size, the liquidity impact is probably minor, as bankers confirmed to this author.

Thus, with a view to rationalising their operations, complying with regulatory exigencies, reducing excessive indebtedness or building liquidity, companies have had to sell under-performing assets in due course, notably assets unrelated to their core line of business. Corporate liquidity would then be used to develop various activities in areas of the companies' proven expertise (as and when such developments became warranted) or to repurchase their own stocks, whenever they were traded much below the company's valuation of their assets. The stock repurchase would raise the stocks' market value and discourage potential corporate raiders. The following policy statement by the chairman of the world's largest oil company gives an eloquent description of that strategy:

> You're all familiar with the fact that precisely because the search for petroleum is so risky, a number of companies have recently sought to strengthen their positions by acquiring already discovered oil and gas reserves from others. Exxon does not rule out such acquisitions. If they make economic sense to us, we're prepared to consider them, and in fact we made several relatively small purchases of reserves in 1984 that seemed to us a good fit with existing operations. But we have not in recent years found major opportunities that in our judgement were in the intersts of our shareholders.
>
> An alternative that we have pursued is to use a portion of our funds to repurchase Exxon stock. This has the virtue of not only increasing the amount of oil and gas represented by each share retained by stockholders but of increasing earnings per share and enhancing shareholders return as well. How long and on what scale we shall continue to repurchase stock will depend on cash flow, investment needs and a variety of other factors. As I have said on a number of occasions, we don't regard this activity as a substitute for capital outlays that meet the return criteria we have established. But under current circumstances it makes good sense.[23]

The repurchase of corporate shares has been either on an ongoing basis or on an *ad hoc* basis. These repurchases return some of the companies' equity to shareholders, raise the market value of shares and contribute in principle to larger earnings per share—assuming, *inter alia*, no contraction in corporate income. Share repurchases would shrink capital and lead to a corporate financial structure 'more highly leveraged with debt'.[24] The year 1984 saw not only a bulge in mega-mergers (see Table 23) but also in mega-repurchases of shares (Table 26).

Table 26 Share repurchases by ten large petroleum companies, 1981–1984 ($ millions)

	1984	1983	1982	1981
Amoco	$1,323	$194	$339	$306
Arco	723	—	—	23
Chevron	—	289	379	—
Exxon	2,672	785	98	—
Mobil	—	11	471	—
Philips	—	—	—	—
Sohio	561	6	—	—
Sun	203	97	106	155
Texaco	1,282	74	81	—
Unocal	—	—	—	—
Total	$6,764	$1,456	$1,474	$484

Source: Bernard J. Picchi, *The Structure of the US Oil Industry: Past and Future*, New York, Salomon Brothers, July 1985, p. 4.

Governments are concerned about excessive lending by banks to finance merger and acquisition transactions or stock repurchases. Heavy reliance on debt and the erosion of their equity base make the companies concerned more vulnerable to economic adversity, such as downturn in earnings or sharp increases in interest rates. A large increase in debt and/or a huge retirement of equity by business enterprises could impair the stability of the financial system. This was expressed by a spokesman for a US central banking authority:

> The Board of Governors, as bank and bank holding company supervisor and lender of last resort, has responsibility, along with the other regulatory agencies, for maintaining the safety and the soundness of financial institutions and markets. In this regard, we are especially concerned about potential financial risks involved with leveraging and with acquisition activity financed with large amounts of debt. Because many mergers and leveraged buy-outs have involved heavy reliance on debt and retirement of existing equity, the surviving firms have had balance sheets that leave them more vulnerable to downturns in earnings or sharp increases in interest rates. When this development occurs, it of course means that the institutions providing the credit may in turn be more exposed to possible loss.[25]

Bank Exposure and Energy Projects

Banks make money by mediating between savers and borrowers, and managing the different maturities of their resources and uses of funds. Commercial bankers are in principle supposed to be risk-averse, in view of their heavy dependence (for the major part of their resources) on depositors' funds, which are largely short term. They should therefore concentrate on short-term or medium-term financing of credit-worthy borrowers and be loath to lend to risky ventures, including untested oil projects (for fear of dry holes, cost overruns, declining proceeds, etc.). Bankers have, however, often failed to abide by these standards of prudence.

General corporate loans obtain more favourable terms than project loans, owing to the guarantees offered by the corporation. Corporate guarantees would secure the lender against the riskiness of an individual project. Given the size and prospects of oil companies, banks have reputedly readily offered attractive conditions to majors, especially in financing their acquisitions of other going concerns, their purchase of minority interests in affiliates or, as mentioned above, the purchase of their own stocks. The terms obtained by major oil corporations were 'comparable with those a top-rated European government could command in the [Euro] market'.[26]

Some banks have been overextended in financing energy projects. Less experienced banks were spurred by intense competition, which led them to neglect insistence on profitable and secure projects, and on adequate spreads to meet risks. Between 1973 and 1980 oil companies and their banks were further lured by rising oil prices, whose ascent was then deemed to be sustainable. With the 1981–85 decline in oil prices, many small energy operators in the United States[27] and several banks in Texas and Oklahoma which were heavily dependent on energy loans sustained considerable losses. This was the case with Penn Square Bank of Oklahoma City, which closed down in 1982. It provoked a spin-off of losses to other banks, including some major ones, notably Continental Illinois and Chase Manhattan. Chase's problem with Penn Square was not critically damaging.[28] The most dramatic impact was on Continental Illinois (the sixth-largest US banking group in 1982), which had bought over $1 billion in energy loans from Penn Square Bank. Losses on these provoked the disaffection of depositors (which included domestic and foreign banks).[29] Continental had in fact depended excessively on short-term and big depositors, including a few depositors from OPEC countries.

The reasons for problem loans are multiple; some of the main ones being:

— poor choice of projects;
— inadequate guarantees;
— unfavourable circumstances in industry, notably the decline of oil prices rendering high-cost projects unprofitable;
— complacency or mismanagement in the handling of soaring big-loan projects.[30]

Trade sources have acknowledged that the biggest challenges to the world financial system in recent years have come from foreign debt, energy loans, real estate and agriculture. According to a leading business journal, the US financial system has suffered more from domestic bad loans (in which energy ranked foremost) than from Third World debts. Its analysts concluded, 'No industry has caused more banks to fall from grace in the 1980s than oil.'[31] It is worth remembering in this context that US banks are required to disclose more information to the public than their counterparts in Europe and Japan. This lack of disclosure in the latter countries should not lead one to conclude that the problems of bad energy loans are problems of US or Canadian banks alone.

Several developing countries have resorted to external borrowing to finance their energy projects. The bulk of such financing between 1975 and 1980 was for electrical power (72 per cent), with oil and gas representing 25 per cent and coal a mere 3 per cent. The largest contribution of publicly guaranteed external financing to developing countries was export-related (39 per cent), followed by commercial institutions (36 per cent) and multilateral development banks (16 per cent). Concessional financing represented the balance of 19 per cent (Table 27). Nearly all the foreign-credit financing was aimed at energy projects with no 'geological' risk relating to the discovery of commercial reserves.[32]

Table 27 Publicly guaranteed external borrowing by developing countries for energy, 1975–1980 ($ billion 1982)

	$	%
Export-related	37.5	39
Financial institutions	35.2	36
Multilateral	15.5	16
Concessional	9.1	9

Source: The World Bank, *The Energy Transition in Developing Countries*, Washington DC, 1983, p. 72.

When borrowing, oil companies prefer to borrow over the long term, spanning the life of the project, and at fixed rates of interest to 'judge more accurately the economics of the project'.[33] Some oil finance, however, cannot be obtained except at floating rates. International financial markets are not normally tapped for 'concessionary' financing.[34] The latter obtain their funding from government contributions drawn from national budgets, and disbursed, either by bilateral agreement or via multilateral agencies.

The apparently attractive terms to borrowers of concessionary financing (subsidised interest rates, long maturities, grace periods, etc.) could hide certain drawbacks. Among these drawbacks, one could mention financing that is 'tied' to the sale of goods or services by donor countries at prices above market rates; alternatively, the goods or services sold could prove inadequate (i.e. inappropriate or inferior in quality). Thus, concessionary financing is not necessarily equivalent to 'cheap' financing. Indeed, some developed industrial countries with limited domestic energy supplies find it in their national interest to offer concessionary financing to energy projects overseas, in the expectation of having preferential assets. This is true of Japan.[35]

Business Climate

The comparative economic attraction of any project is considerably influenced by existing laws and government action. Financial-fiscal incentives and operational facilities can enhance the viability of an energy project which would otherwise remain uncompetitive. On the other hand, exacting terms for operators (e.g. large signature bonuses, high royalty and tax rates, heavy investment obligations, or operational constraints, etc.) can decrease the attraction of a venture with respect to prospective income and can consequently deter potential entrepreneurs and investors.

The transnational petroleum industry generally perceives political risks attached to developing countries to be of such magnitude as to deter exploration. Thus in 1981, some 80 per cent of the industry's exploration expenditure was devoted to developed countries, where the average cost per barrel of oil and gas discovered was about four times that of the cost in the oil-importing developing countries. More specifically, the cost per added barrel of oil or gas between 1974 and 1983 was $4.48 for developed countries, $1.15 for oil-importing developing countries and $0.24 for oil-exporting developing countries.[36]

Analysts have noted that the fiscal systems of some transnational corporations' home countries (the home countries happen to be principally the developed countries) do not give foreign exploration or investment activities 'parity of treatment' with domestic operations. These analysts also argue that the absence of tax credit on the foreign operations of transnational corporations have favoured the development of the energy potential in developed countries, at the expense of developing countries.[37] The bias in favour of the home country is most conspicuous in the United States where, at the individual level, there is the possibility of deducting from personal income a high share of exploration costs in the year incurred; at the corporate level, there is a depletion allowance shielding 22 per cent of gross domestic petroleum income from taxation, and the permission for deducting losses sustained in domestic petroleum operations from domestic non-petroleum income. The effective corporate tax rate on petroleum income from domestic petroleum operations in the United States was estimated at less than 15 per cent.[38]

Other governments, needless to say, are free to match US terms. More generally, a measure of flexibility in fiscal terms and in operational requirements demanded from investors is needed to enable the latter to obtain adequate profits—depending on the size of discovered reserves and the cost of producing from them. Thus, exploration for petroleum in untested areas warrants a moderation in (or even the dispensing with) rentals, bonuses and other payments, as well as a reduction in royalty and fiscal charges for the pioneering company or companies.

Moreover, the insistence on heavy work requirements over a short time-span can prove counter-productive. Companies need time to assess geological data in virgin areas before committing additional expenditure or opting to relinquish exploration areas without penalty.[39] An oil company's interest in getting into a new exploration venture is not prompted by the expectation of 'normal' return, equivalent to the average performance of the energy industry or similar risk industries. It is prompted by the expectation of an 'above normal' return, sufficiently attractive to ensure that the yields of possible exploration successes compensate for likely losses.

FINANCING TECHNIQUES

The mobilisation of additional financial and technical resources aimed at the efficient development of commercial energy world-wide appears desirable for a number of reasons:

- there is a need to expand the energy base since the so-called 'glut' of 1982–85 will be re-absorbed once economic recovery picks up momentum;
- much of the energy used in the twentieth century has been derived from depletable resources (like coal and other hydrocarbons); this demonstrates the need to expand and diversify sources of energy in order to make up for their exhaustion;
- the energy potential of developing countries, compared with that of developed countries, has so far been inadequately explored and developed.

A few developing countries have been loath to accept foreign direct investments, perceived to be at variance with their economic ideologies and their wish to maintain full effective control over their national economies. These countries' preference is for debt or equity financing which does not carry managerial control of domestic ventures.

It is well established, however, that reliance on debt-financing could undermine the viability of oil exploration and development projects. This is due to the fact that large sums must be spent in the expectation of an eventual return which could take several years to materialise, if at all. Even if reserves are discovered, the size of the return is uncertain, owing to the risks involved in geology (i.e. regarding the discovery of commercial quantities), in the market (i.e. the adequacy of sales profits), and to political risks (i.e. the stability of contractual relations). If such uncertain future benefits were to be further saddled with fixed financial charges (i.e. the payment of interest and the reimbursement of principal) at pre-agreed dates and levels, then the economic standing of the oil or gas projects would be further strained. This can be compared with an equity-financing situation, which does not call for these scheduled payments, or for remunerating or redeeming capital should the project prove unproductive.[40] Financial arrangements have, therefore, to be tailored to the conditions of the projects being financed. Some debts techniques are reviewed hereafter.

Commodity-Linked Debt

Loans, if and when resorted to, should be arranged such that amortisation and interest payments start once the project begins to yield income. Thus, in 1972, BP contracted a private-bank loan of £180 million and $468 million for the development of the Forties Field in the North Sea. Repayment was contingent on the production of oil, and was fixed at a level not to exceed profits from the field. It was agreed that any interruption in the flow of oil would automatically lead to an interruption in debt repayments. Other independent companies with moderate financial strength (i.e. with low-equity base in relation to total liabilities) had to entice lenders by assuring them, not only with interest payment and reimbursement of principal, but also with a share of profits from the fields financed by the credit in question. This approach was notably used in the oil development of the British North Sea.[41]

'Commodity price-linked' credit techniques would be appropriate for those host countries which would not accept or obtain foreign direct investments. Such credits would have their yields and/or redemption value linked to the price of oil. The debtor thus transfers the market risk (i.e. price) to the creditor, who in turn expects to be duly rewarded for his assumption of that risk.

Mexico has resorted to petro-bonds, where both principal and interest are linked to the price of Mexican oil, thus protecting the borrower against the decline in the price of its oil (expressed in US dollars) and enabling the lender to share in an eventual price increase. The lender, who is providing hard currency (i.e. US dollars) funding, naturally expects reimbursement in the same currency.[42]

A developing country must be well rated by creditors if it is to have the opportunity of financing its exploration and development programmes by oil-linked debt instruments in the international financial markets.

Zero-coupon Bonds or Notes

Another financial instrument which could prove particularly attractive to energy projects is the zero-coupon bond or note.[43] The investor derives his interest and recovers his capital at the term of the loan. The absence of regular interest payments can be a relief to the borrower who does not expect his project to yield income for several years. The zero-coupon might also have an appeal for investors who want to have their interest yield automatically capitalised and re-lent to the borrower, and who expect some fiscal advantages as a result. Nevertheless, some tax codes (e.g. in the United States) require annual tax payments on imputed annual interest.

Co-financing Linked to Escrow Accounts

Project finance by banks does not necessarily enjoy the blanket guarantee of the corporation(s) exploiting the project or that of the host government concerned. Creditors, however, can obtain some protection on such project financing by a number of means, including:

— Insurance against default or political risks provided by private or official agencies of countries exporting capital goods or services for the project under consideration.
— Co-financing arrangements with a cross-default clause feature, serving to inhibit debtors and their governments from declaring default *vis-à-vis* any single creditor, lest the debtor be confronted by general disaffection from all the other creditors.

To secure and entice the financing of energy projects further, the proceeds of such projects, once they come onstream, can be geared to escrow accounts. This approach generally appeals to financiers of export-orientated projects. It is often a necessity in countries which are shunned by creditors. Creditors can then be secured by the prospect of having the project's 'export proceeds' designated specifically to satisfy on a first-priority basis the claims of the project's creditors (i.e. before any other use is made of these proceeds).

Revolving Fund for Exploration

Oil business, it has been established, is an extremely risky activity. Pre-investment geological surveys and feasibility studies can minimise the risk of expensive mistakes but several developing countries often lack the financial and technical means to carry out preliminary investigations. A variety of contractual arrangements have evolved over the years which effectively allow the host country to gain information on the extent and nature of its potential subsoil minerals by entering into agreement with major transnational, state-owned, or independent companies. Normally, the company accepts the exploration risk in return for participation in the ultimate benefits of the project, and in accordance with a predetermined formula. However, should host governments not reach agreement with commercial companies on exploration and production terms, or should host governments refuse foreign companies a direct role in the development of their natural resources, alternative avenues should be looked into. In this respect, one could propose the creation of an International Revolving Fund (IRF), with a sufficiently large financial base. The IRF's resources would be used to underwrite pre-investment exploration and feasibility studies made to ascertain the attractiveness of energy projects. Once these pre-investment investigations had led to the floating of a commercially viable project, the beneficiary company or country concerned would have to pay pre-agreed fees, the value of which could be made to vary with costs and/or profits. If, however, the host country of a productive exploration programme were to find itself unable to pay the agreed fee, the IRF could take up participation in the venture responsible for the commercial exploitation of the project (to the extent of the value of the fee) for an interim period, until it found one or more suitable buyers for such participation.

It would appear reasonable to vary the fee according to commercial quantities produced, prices and/or profit margins. In such arrangements the host country and its beneficiary company bear no cost for unproductive ventures. Even producing ventures faced with declining prices or escalating costs will be protected against payment of such a fee if no profits accrue.

As compared with national or regional institutions, the IRF can spread risks, thus compensating losses incurred in one country, region or branch of activity with profits obtained in other areas. The IRF need not be limited to large or export-orientated projects. Small projects can also be economically viable and socially helpful to the domestic economy. Once the feasibility/exploration stage has been completed, the IRF need not be involved in the development and/or commercialisation of energy projects. These activities would then be left entirely to energy enterprises, national, transnational or independent.

Aided by appropriate investment banking advice, the IRF would be better equipped than a developing-country agency to introduce appropriate energy projects of that country to international financial markets. It could thus assist in floating financial packages for countries which have been shunned, largely out of ignorance, or which have been subjected to unreasonable terms.

To entice private as well as public financial support, the IRF should be concerned with achieving attractive returns commensurate with risks encountered. These returns should serve to cover expenses (including eventual losses) and to generate profits to be used not only for reasonable dividends but also for expanding the financial and operational base of the IRF.

Secondary Market in Financial Assets

Financing an industry is more attractive to an institution the greater opportunity there is for that institution to shed into the secondary market individual assets from their portfolio. A secondary market encourages lenders and investors since the latter are assured liquidity by the possibility of disposing of some of their assets when they are in need, or when they are contemplating new lending or investment opportunities. Lenders would also have the opportunity of disposing of certain assets for the sake of diversification. Low-yield or delinquent loans have to be discounted (with due concern for the reward–risk balance) to attract potential investors. To facilitate the mobilisation of loans, these can be further segmented and adapted to the need of the market-takers. Various elements of the loans can be re-tailored to suit potential investors: maturity periods of principal, interest formulas, multicurrency options, guarantees, etc.[44] Borrowers would more readily find financiers if their creditors could obtain the kind of flexibility mentioned above.

In conclusion, we shall again raise the question of how far actors in the oil industry have affected international financial markets. The evidence does not support the contention that capital-surplus OPEC countries have had a destabilising impact.

What about the capacity of the international financial system to respond to energy crises? Despite certain mishaps encountered by individual borrowers and lenders, financial institutions have generally demonstrated considerable resilience, aided, admittedly, by public authorities. It can nevertheless be argued that the oil challenge has stimulated innovation in the development of lending techniques by the international financial markets. Substantial financial resources are bound to be mobilised in the direction of the oil sector, owing basically to three factors:

— the continued importance of oil in the world energy market and in world economic growth;
— the rising investment costs of discovering, developing and commercialising oil products;
— the large volume of international debt-financing resorted to by various private and state-owned petroleum enterprises.

NOTES

1. See, for example, James D. Hamilton, 'Historical Causes of Postwar Oil Shocks and Recessions', *The Energy Journal*, **6**, 1, 1985, pp. 97–116.
2. For example, *BNOC, the British National Oil Corporation*, London, June 1984.
3. R. T. Mcnamar, Deputy Secretary for the US Treasury, 'The Portfolio Shift Theory of Exchange Rate Movements', Davos Symposium of the EMF Foundation, Switzerland, 1 February 1985, typescript.
4. See, for example, Kevin Muehring, 'Arab banking—the new look in Arab deal-making', *Institutional Investor*, New York, June 1985, pp. 193–216.
5. See also Samuel L. Hayes and Zuhayr Mikdashi, *Middle East Capital Flows*, Working Paper, Harvard Business School, Boston, 1985.

6. In *International Debt and the Stability of the World Economy*, Institute for International Economics, Washington DC, September 1983, p. 21–5.
7. 'AEA Meeting Highlights Conflicting Views over Issues of Debt and Exchange Rates', *IMF Survey*, Washington DC, 7 January 1985, p. 11.
8. Bank for International Settlements, *Fifty-Fourth Annual Report*, Basle, 18 June 1984, p. 171.
9. See 'Latin America: the other side of debt', *The Economist*, London, 23 June 1984, p. 75.
10. George J. Clark, Executive Vice President, Citicorp, in *World Debt Situation*, Hearing before the Subcommittee on International Economic Policy, US Senate, Washington DC, 27 September 1982, p. 10.
11. *54th Annual Report*, p. 171. For similar interpretations, see R. T. Mcnamar, Deputy Secretary for the US Treasury, *The International Debt Problem: Recent Progress and Future Ideas*, Davos Symposium of the EMF Foundation, Switzerland, 30 January 1984; and Karin Lissakers, 'Dateline Wall Street: Faustian Finance', *Foreign Policy*, New York, Summer 1983, pp. 160–75.
12. Dome Petroleum Ltd, *June 30, 1984* and *March 31, 1985*, *Interim Reports*, pp. 13 and 8, respectively.
13. See *The Economist*, 22 September 1984, p. 99, and 18 August 1984, p. 62.
14. Testimony and response to questions by James E. Lee, Chairman of the Board and Chief Executive Officer of Gulf Corp. (pp. 103–11 and 152–5), and of Boone T. Pickens, jr., chairman and chief executive officer of Mesa Petroleum Co., (pp. 163–91) in US Senate Committee on the Judiciary, *Oil Merger Activity*, Washington DC, 15 March 1984.
15. Edfred L. Shannon, jr., 'Merger Activities in the United States Petroleum Industry', in *Dinar*, The Commercial Bank of Kuwait, 1985.
16. 'Buying Cheap Reserves on Wall Street', in Arthur Andersen and Co., *Oil and Gas Reserve Disclosures, 1980–1983*, Houston, Texas, 1984, p. S-15.
17. Chairman Robert O. Anderson and President William F. Kieschnick in Atlantic Richfield Company, *1984 Annual Report*, Los Angeles, 1985, p. 3. See also 'Arco enters oil's new era: the majors are slimming down before the raiders do it for them', *Business Week*, New York, 13 May 1985, pp. 16–17.
18. Mobil *Annual Report 1984*, pp. 2 and 24.
19. See *Financial Times*, 5 April 1984, p. 29.
20. *Business Week*, New York, 8 October 1984, p. 23.
21. See Jack Willoughby, 'A mixed blessing', *Forbes*, New York, 19 November 1984, pp. 44 and 46.
22. *Oil Merger Activity*, op. cit., pp. 146 and 163; and Chevron Corporation, *1984 Annual Report*, San Francisco, 1985, p. 39.
23. C. C. Garvin, '1985 Annual Meeting of Shareholders', *Exxon News*, June 1985, pp. 2–3.
24. For example, William F. Kiechnick, President and Chief Executive Officer of Atlantic Richfield Company, *1985 First Quarter Report and Summary of Annual Meeting*, Los Angeles, California, p. 14.
25. Statement by Preston Martin, Vice Chairman, Board of Governors of the Federal Reserve System, before the Subcommittee on Securities of the Committee on Banking, Housing and Urban Affairs, US Senate, 4 April, 1985, reproduced in *Federal Reserve Bulletin*, Washington DC, June 1985, p. 418.
26. P. Montagnon, 'Cash-rich banks offer good terms to corporations', *Financial Times*, 5 March 1984, p. 17.
27. Estimated at one-fourth of the country's 20,000 wildcatters, *Business Week*, New York, 1 July 1985, pp. 36–7.
28. 'Bank Insolvent as Drilling Declines', *Petroleum Economist*, London, September 1984, p. 385 and 'Chase's battle to catch-up', *Business Week*, New York, 9 April 1984, pp. 54–5.
29. For an extensive account, see 'How Not to Run a Money-Center Bank: The Report of

Continental Illinois Corporation's Special Litigation Committee', *International Currency Review*, London, September 1984, pp. 14–29.
30. Geoffrey Smith, 'Who was watching the store?', *Forbes*, New York, 30 July 1984, p. 37.
31. 'Behind the banking turmoil: bad loans at home—not in the Third World—are shaking the system', *Business Week*, New York, 29 October 1984, p. 59.
32. The World Bank, *The Energy Transition in Developing Countries*, Washington DC, 1983.
33. British Petroleum Co. Ltd., *Our Industry Petroleum*, London 1977, p. 41.
34. The Committee on Aid for Development of the OECD defines 'aid' as any finance which contains a subsidy element equivalent to at least one quarter of the aid's face value. The subsidy element is calculated thus: 1 *minus* the ratio of (discounted value of the repayment of the loan plus interest payments *divided* by the face value of the loan). The discount rate used is that of the appropriate market rate of interest.
35. 'Eximbank Finance for Energy Resources Development Projects', by M. Nakayama, Director-General of Research Institute of Overseas Investment, The Export–Import Bank of Japan, Tokyo, 1984, p. 3.
36. C. R. Blitzer, *et al.*, *Oil Exploration in the Developing Countries: Poor Geology or Poor Contracts?*, Energy Laboratory, MIT, Cambridge, Mass., 1985, cited by *Petroleum Economist*, London, May 1985, pp. 154–5.
37. See Keith Palmer, 'Private-sector Petroleum Exploration in Developing Countries', *Finance and Development*, IMF–World Bank, Washington DC, March 1983, pp. 36–7.
38. Ibid., p. 37.
39. See, for example, proceedings of *United Nations Symposium on Financing of Petroleum Exploration and Development in Developing Countries*, Athens, 22–27 April 1985, including T. P. Brennard, Shell, 'The Role of the Private Sector in Petroleum Exploration'; United Nations Centre on Transnational Corporations, *Main Features and Trends in Petroleum and Mining Agreements*, New York, 1983 (ST/CTC/29); and Raymond F. Mikesell, *Petroleum Company Operations and Agreements in the Developing Countries*, Resources for the Future, Washington DC, 1984.
40. See, for example, T. Agmon, Donald R. Lessard and J. L. Paddock, 'Appropriate Financing for Petroleum Development in Developing Countries', *The Energy Journal*, Cambridge, Mass., **5**, 3, 1984, pp. 27–40.
41. *Our Industry Petroleum*, op. cit., pp. 42–3.
42. See W. Chislett and P. Montagnon, 'Mexican bank makes record petrobond issue', *Financial Times*, 29 April 1983, p. 21.
43. For example, Exxon's US $1.8 billion twenty-year notes issued on 15 November 1984: *The Economist*, London, 17 November 1984, p. 38.
44. See, for example, Michael Bray, 'Developing a Secondary Market in Loan Assets', *International Financial Law Review*, London,October 1984, pp. 22–5.

6 Energy and Development Issues

Oil-exporting countries generally considered that the surge in oil revenues in the 1970s was a welcome opportunity for rapid growth and prosperity. These revenues came to supplement low levels of domestic income and savings. Some major developmental challenges encountered by oil-exporting countries are reflected on in this chapter.

PROGRESS AND OBJECTIVES OF PRODUCING OIL RESERVES

Conventional concepts and computational methods of national income accounting offer only approximate measures of welfare or progress. They suffer from a number of shortcomings, chief among which are:[1]

- Lack of proper assessment of the depletion of exhaustible, non-renewable mineral assets, of the erosion of arable land, and of the pollution of air, water and other natural resources. By omitting to depreciate sufficiently for loss in the quantity or quality of natural resources, GNP figures for the countries concerned are overstated.
- National production may be destined for either military or civilian use—each with a different social impact. Military expenditure competes with basic civilian requirements, such as better schooling, better medical services, safe water, cheap and convenient transport, salubrious habitat.
- Inadequate accounting for the distribution of income and leisure. Income concentration among the few often proves a source of tension. By comparison, the assurance of basic necessities to the deprived sections of society and the avoidance of wide disparities in wealth and opportunity are generally conducive to enhancing national welfare and harmony.
- Social costs are inadequately accounted for; for example, the social costs of accidents, pollution and transport. Thus, additional expenditure to cover accidents, pollution or travel costs for longer journeys to work all raise the statistical gross national product (GNP) but not the standard of living, i.e. quality of life.
- Factors of efficiency of public institutions and of participation in economic and political decisions, job security, the opportunity for

advancement, the degree of harmony among and within social groups and similar qualitative variables are not readily recordable in statistical form.[2]

— The exclusion of certain economic activities (e.g. housework), while other activities elude official statistics, such as the 'black' economy, subsistence production (mostly in developing countries), and non-marketed incomes.

— Conventional national income accounting gives rise to problems with respect to comparability of figures across years and across nations. The construction of deflation indices to express the GNP of a country in constant prices is a challenging exercise: goods and services, whose price is included in a given index, vary over time with respect to their weight in the national product of an economy and with respect to their quality. Moreover, GNP figures of various countries are not readily comparable because of problems of making appropriate exchange-rate conversions.

Nevertheless, national development programmes and strategies generally strive towards five main and interrelated goals: satisfactory, self-sustained and balanced rates of growth; diversified and versatile economic structure in order to reduce economic vulnerability to unexpected shocks; judicious development of national resources (the human and the physical); financial stability; and survival as a nation state. The latter is admittedly an economic as well as a geopolitical–military objective, and often takes precedence.

The availability of financial resources was not a constraint for a number of oil-exporting developing countries in the 1970s. Many, however, have encountered challenges in investing funds productively. A trade-off exists between, on the one hand, keeping oil in the ground and, on the other hand, selling it to obtain the requisite financial resources. An equilibrium point could be reached at the level of production where expected appreciation of oil in the ground is *equal* to expected return on investments (whether domestic or foreign investments), taking due account of socio-political benefits or costs at domestic, regional or international levels. To illustrate, an increase in oil production by a country may not be viewed positively by different national groups or sections of a country's governing elite anxious to conserve oil resources for future needs. Yet such an increase may be resorted to by the government concerned either to keep or win friends among oil-importing countries concerned about adequate oil supplies at reasonable prices.

Put in technical terms, the profitability of keeping oil in the ground depends entirely on whether the rate of appreciation in real terms of the oil in the ground is greater than the 'social' discount rate. The latter represents the return to society from using oil proceeds in various economic activities. The greater a society's urgency for obtaining funds for various uses, the higher the social discount rate and, consequently, the greater the pressure to produce oil in order to obtain oil revenues. Revenue is needed to run the state, to ensure security and to underwrite essential development projects.

INDUSTRIALISING PETROLEUM

Countries' motives for industrialising their natural resources, including petroleum, are varied. Depending on the circumstances of the country concerned, emphasis is laid on one or more factors:

- diversification of revenue sources;
- value added and net social gains obtaining from the process of industrialisation;
- employment, notably in countries with large populations and disguised or open unemployment;
- forward and backward linkages in the domestic economy;
- reducing 'excessive dependence' on imports;
- generation of additional foreign-exchange earnings.

Petroleum can thus provide domestic opportunities for the development of 'downstream' operations such as refining and petrochemical plants. In exploiting these opportunities, the country's industrial base can expand using, for example, abundant hydrocarbons as a launching pad for the development of related new business activities. It was estimated that in 1983 about 28 per cent of the associated gas produced in the OPEC countries was flared, in the absence of laws, regulations or incentives to avoid waste and to invest in industries capable of transforming gas into a useful source of energy or a feedstock for chemical products (such as chemical fertilizer and other petrochemicals).[3]

One of these ironic situations is to see associated gas being flared while nearby wood is being cut for fuel at the risk of deforestation, and agricultural land is left underfertilised. This has been the case in Nigeria. Its flared gas was estimated in 1983 at 13 billion cubic metres, equivalent in value to some $4.4 billion, or about half of Nigeria's oil exports.[4] Yet not all associated gas can be productively used. This is especially the case where oil-fields are small and widely scattered, thus making their development as well as the transmission of associated gas too costly.[5] In 1985, on the basis of a 1979 government decree, Nigeria started enforcing a penalty of two US cents per 1,000 cubic feet of 'unnecessary gas flaring by oil companies'.[6]

Often under the pressure of public opinion, governments of oil states have launched various projects to harness hitherto flared gas; for example, through liquefaction plants, the petrochemical and fertiliser industries, the establishment of energy-intensive industries, large-scale export projects, the use of gas locally in power stations and desalination plants, and the reinjection of gas to maintain reservoir pressures.

In industrialising their petroleum reserves, developing countries need to have continuous access to up-to-date technology and dependable external market outlets. This has often prompted the host countries concerned to elicit the partnership of major transnational petroleum and chemical groups. Thus Saudi Arabia enticed majors to join the state-controlled Saudi Basic Industries Corporation (SABIC) in launching joint ventures. Besides abundant cheap raw materials, the foreign partners benefit from interest-free loans, with tax holidays until the joint projects show profits. If the return on equity investment (ROE) is 10 per cent or below, the interest rate is 3 per cent; it rises to 4 per cent for a ROE of between 10 and 15 per cent, to 5 per cent for a ROE of between 15 and 20 per cent, and to 6 per cent for a ROE of 20 per cent or more.[7]

Notwithstanding lower-cost inputs (notably oil and gas), the challenges faced by oil-exporting countries in entering the petrochemical sector remain considerable. They encounter various trade barriers in the major markets of the world; that is, those of the industrial countries. In recessionary periods the industrial countries are likely to maintain or strengthen measures aimed at protecting their own industries. Thus, in June 1984, the countries of the European Economic Community (EEC) imposed tariffs and other protectionist measures on basic and intermediate petrochemicals (methanol, ethylene, ammonia and others) imported from OAPEC countries (particularly Saudi Arabia). The tariffs, ranging from 13.5 to 20.0 per cent, are seen to undermine the economic viability of oil exporters' petrochemical ventures and thereby frustrate their chances of industrialising and developing. In 1983 Saudi imports from the EEC reached $20 billion, twice as much as Saudi exports to the EEC. Some 75 per cent of imported Saudi merchandise was totally exempt from customs duty, with the balance imposed at a mere 3 per cent.[8] The Saudis are currently eligible (1985) for a nominal duty-free quota of ECUs256,300 ($187,500). Their aim is to reduce the EEC common customs tariff on GCC petrochemical products to the 4–7 per cent range.[9]

The EEC attitude is based on a complex set of factors. Petrochemicals are considered to be 'strategic' products by the EEC. Moreover, faced with an import lobby of European petrochemical producers with less efficient plants threatened with closure in a stagnating market, a number of EEC countries supported the imposition of tariffs (i.e. West Germany, Italy and the Netherlands).[10]

The location of processing or refining industries next to the sources of raw materials is not necessarily a zero-sum game. While it can prove beneficial to producer countries, it can also be beneficial to consumer countries, particularly when it:

— reduces congestion in ports and/or suitable sites in densely populated importing countries such as Japan;
— reduces the emission of pollutants;

— reduces global energy consumption;
— stimulates the economic development of raw material exporters, with a spill-over impact on various enterprises in industrial countries interested in supplying to, or contracting in, developing countries. These enterprises include, among others, engineering and equipment groups interested in downstream operations and related processes and technologies.

Arab Gulf states have espoused development strategies favouring capital and energy-intensive industries. Their large petroleum reserves, their small populations and their limited agricultural potential explain this preference. Other oil-producing countries with large populations have been, by comparison, very much concerned with industries capable of generating significant employment.[11]

Industrialisation in oil-exporting countries has suffered from a number of constraints. The four main ones are:

— trade or financial impediments imposed by major trading partners;
— lack of domestic infrastructure and supporting amenities;
— limited domestic markets (in terms of population or purchasing power);
— absence of industrial tradition or know-how, and scarcity of indigenous managerial talent.

Petroleum-based industrialisation in oil-exporting countries is criticised on the grounds that it does not promote diversification, to the extent that this type of industrialisation accentuates dependence on the petroleum sector. Whilst this argument could prove valid under certain conditions, oil-exporting countries have undeniable advantages in using petroleum for industrialisation. In particular, gas associated with crude-oil extraction (representing about 12 per cent of the energy content of a unit of extracted crude oil) could be used in petrochemical industries instead of being flared. Moreover, the social benefits of petroleum industrialisation in the Gulf countries, in the absence of other domestic opportunities, are reckoned to be positive on balance. As one analyst has said, 'Downstream industries, apart from being consistent with these countries' endowment of natural resources, provide excellent laboratories where workers, engineers, technicians, managers, etc. can acquire useful skills'.[12]

OIL PRODUCERS AND FOOD PRODUCTION

Several developed economies, such as the United States and Western Europe, have shown that their advancement in the industrial and service sectors cannot be dissociated from a concomitant advancement in the agricultural/rural sector. A number of oil-exporting economies have, in the wake of the oil boom, neglected and virtually abandoned their agriculture, as was admitted by a Nigerian minister. He noted that, as a result, Nigeria has become a mono-product economy and 'very vulnerable to negative shocks, even mere tremors in the global oil market'.[13]

Several factors have been responsible for the atrophy of the agricultural sector, in particular:

— a substantial overvaluation of the national currency, along with the fixing of low prices on basic food products, helped to encourage food imports at the expense of domestic agricultural production;
— political favouritism for a dominant affluent minority, contrasting with the poverty and hunger of the rural majority and of those living in urban squalor;
— disbursement of oil revenues was largely biased in favour of urban areas; this has drained farm-workers away from rural areas, lured by higher wages and better amenities;
— neglect of infrastructual investments in rural areas, for example in transport, power, storage-producing facilities, education and training institutions and medication.[14]

Other oil-exporting countries with a negligible or moderate agricultural base have pushed for the development of that sector through massive subsidies. Their primary goal is a certain measure of food security. Their major means of satisfying that objective is oil revenues. This is notably true of a few Arab countries in the Gulf, and Libya. The policy of these countries has its origin in 'former secretary of state Henry A. Kissinger's suggestion in 1973 following the first oil shock that [US] food could be used as a weapon against the oil producers'.[15]

In the quest for food security, the Saudi Government allocated $21 billion for agriculture and irrigation in its 1980–85 development plan. The Saudi goal is food self-sufficiency, in the absence of reliable suppliers.[16] Direct subsidies apply to such items as financing, water, electricity, fuel, agricultural machinery and equipment, animal feed, fertilisers, agricultural crops, saplings, transport costs of imported high-breed livestock, and training of Saudi personnel.

Food security could be improved not only by subsidising domestic agriculture but also by setting up a strategic stockpile to cover a few months' consumption. The US secretary for agriculture, John R. Block, thought it 'crazy' to see Saudis spending billions on developing agriculture in a physically hostile environment while the United States can supply much cheaper farm products and is taking farm-land out of production. The Saudis are concerned about a possible food embargo, as the United States has practised *vis-à-vis* the Soviet Union and Iran.[17] The Saudi Government itself resorted to an oil embargo on the United States late in 1973 in the hope that the latter would refrain from helping Israel until Israel had withdrawn from the Arab territories it had occupied following the 1967 hostilities. However, in general, embargoes have proved of no avail. Accordingly, since embargoes are hard to enforce, and in view of the small size of Saudi food imports, one could query the wisdom of a self-sufficiency policy regardless of cost.

To illustrate, the Saudi Government has been offering its wheat growers a procurement price of $1,000 a ton, at least seven times the current commercial price, in addition to other subsidies covering, *inter alia*, land grants, free-interest loans and subsidised water. From a mere 8,000 tons in 1975, the wheat harvest of 1984 had reached 1.3 million tons, with a small surplus for export. Three problems are, however, of major concern to the authorities:

— the decline in the water table of wells in a desert country like Saudi Arabia (with 3 inches of annual rainfall);
— the high dependence by farms on an expatriate work-force;
— the high support cost of domestic food production.[18]

Libya has been equally concerned about food security. In 1985 its leader Muammar al-Gadaffi gave top priority to the construction of a so-called man-made river. Underground water would be extracted from the Sahara Desert 350 miles south of the Mediterranean coast, and conveyed north for agricultural and urban use. It was reported that the system would have an ultimate maximum capacity of some 58 million barrels of water daily, capable of irrigating nearly half a million acres of land over a period exceeding fifty years. This grandiose scheme, with up to 2,500 miles of pipeline, would eventually cost $25 billion—with the first phase involving 250 water wells at the cost of $3.6 billion, for completion by 1989.[19]

ENERGY FROM AGRICULTURE

The food–energy relationship is two-directional: energy is an input in agricultural production as well as an output of agriculture. Energy as input provides power (mechanical work, change in temperature, light, etc.) for farms and the agri-industry. Some energy products (notably oil and gas) are also raw materials for other inputs essential to agriculture, such as chemicals, fertilisers, pesticides, fungicides, herbicides, health products, animal feedstock, irrigation tubes—to mention a few. It is important to note that disease and malnutrition are serious impediments to higher productivity in agriculture.[20] Human beings and the animals they use can be disabled by lack of proper protection. Millions of people suffer from diseases which can be cured or prevented by appropriate, reasonably priced and readily available medicines and chemicals—as well as by better education. The debilitating impact of such diseases on scores of millions of rural and urban people is well known.[21]

Expectations of continued increases in the price of energy have elicited more research into energy-efficient and science-intensive food production. This production approach would call, *inter alia*, on 'biological nitrogen fixation, genetic improvements and improvements in photo-synthetic efficiency'.[22] Higher energy prices have furthermore promoted the development of bio-energy, notably the so-called 'green alcohol' or ethanol, derived from sugar-cane, manioc, hay, feed grains and forests.

Raw materials for the production of ethanol are mostly plant products which contain sugar or carbohydrates—excluding, that is, synthetic production of ethanol from ethylene, a petroleum derivative. Following the oil price increases of the 1970s, the US Government sought to diversify its sources of automotive fuel, outside the natural crude-oil sector. Accordingly, as we have briefly mentioned (p. 83), it offered various incentives to encourage the production and consumption of agricultural ethanol. Production rose from practically zero in 1978 to about 210 million gallons in 1982 (representing about one-fifth of 1 per cent of US gasoline consumption). It is reckoned that ethanol contains one-third less energy than gasoline, with comparable emission levels and a high-octane rating.[23] It has been established that:

> without the federal gasoline tax exemption, the US domestic fuel ethanol industry would not exist. Indeed, fuel distributors and industry financial analysts agreed that only with substantial government subsidies [$0.50 a gallon of federal subsidy and about $0.40 a gallon of state subsidy in many states, out of a retail price of about $1.65 in March 1984] has ethanol been able to enter the national automotive fuel stream and thereby attract potential producers to the business.[24]

Following on the oil price increases of the 1970s, Brazil, like other energy-importing countries, has sought the development of alternative energy sources—oil and non-oil. This was central to its policy of energy diversification and conservation. Thus, Brazil's state-owned Petrobas's search for domestic oil (directly or with the recent enlistment of transnational majors) led to some moderate discoveries. Since the 1970s, Petrobas has obtained leases in various developing countries: Algeria, Angola, China, Guatemala, Iraq, India and South Yemen. Its greatest oil find was in Iraq, but it sold its exploitation rights there early in 1980.[25]

Probably one of Brazil's most ambitious energy programmes is deriving domestic ethanol from sugar cane. This was done largely by subsidised credit (estimated at about US$3.7 billion annually), mostly financed by a levy on gasoline (about $1.6 per gallon in 1981). The price of alcohol fuel was maintained at 65 per cent of the price of gasoline, and taxes were lowered on alcohol-powered cars.[26] Brazil's alcohol production rose from 580 million litres in 1975 to over 4 billion litres in 1981, with a target production of about 8 billion litres (equivalent to 50.3 million barrels of oil) set for 1985.[27] Among the advantages of fuel alcohol, in comparison with gasoline, are clean-burning, its production from a renewable resource and its relative labour-intensity; moreover, the location of alcohol distilleries close to crop areas vitalises the rural economy. However, there are four main drawbacks to fuel alcohol: its corrosive impact on metal; the polluting effluents of alcohol distilleries;[28] the competition of energy from agriculture with food crops;[29] its lower energy content and much higher production cost compared with gasoline.[30]

On the food–energy link, the Brazilian example is instructive. Indeed, the output of staple foods in that country (i.e. rice, beans, corn, cassava and potatoes) fell sharply by over 25 per cent per capita between 1977 and 1983, while that of sugar cane increased by 57 per cent. The hardest hit in food shortages were the poor, who represent two-thirds of the Brazilian population, with a food intake below the 2,480 calories adequacy level (according to the norms of the Food and Agriculture Organisation, FAO).[31]

In the poor developing countries, the priority for food is fundamental. The use of energy crops to fuel vehicles can only be sustained by affluent communities—which happen to be the few. Should there be a substantial reduction in food crops in favour of fuel crops, a majority of the world's population is likely to be faced with rising food prices, implying an intolerable drain on their income, or actual shortages of food to meet their nutritional requirements.[32] While industrial countries can afford to import their food requirements, this is not the case for many developing countries. Various estimates indicate that up to 1.3 billion human beings suffered from malnutrition in the early 1980s, with some 450 million belonging to the very critical category.[33] The plight of that huge section of humanity has seldom caught the active and continuous concern of affluent individuals and communities, which have come to the rescue in somewhat intermittent fashion—mostly at the height of disasters or wars. Often, their aid has proved both ephemeral and uncoordinated.

For the developing countries which can afford food imports, it is remarkable to note that their imports of two leading staple products come mostly from the industrial countries; the latter supply the former's domestic requirements of more than 50 per cent of cereals and more than three-quarters of the milk-product requirements of the developing countries.[34]

Humanity's greatest challenge is that world food production must be at least doubled by the year 2000 to meet minimum global nutritional requirements.[35] Accordingly, fuel from food (cereals, root crops or sugar-cane) should be viewed as an expedient for a transitory period accessible to countries with vast arable lands. Bio-energy could also be derived from the cellulose produced from timber. According to an informed source, 'US forests and crop residues could provide huge amounts of alcohol—up to 39 billion gallons annually by the year 2000—far in excess of alcohol production achievable through the use of grain'.[36] Added to this, large areas unsuitable for crops, if properly forested and managed, could produce wood which would meet the cooking and heating requirements of large numbers in developing areas.[37]

Needless to say, human ingenuity is needed to promote the commercial production of fuel from sources which do not compete with nutrition. One substitute for ethanol is methanol, which can be produced from coal or wood. Ideally, the technology for producing energy from non-polluting renewable resources (e.g. sun, bio-gas, wind or water power) merits further development.

DEVELOPMENTAL PROBLEMS

A few oil-exporting developing countries were in a uniquely favourable financial situation over the 1970s. Over that period they derived large oil revenues, coupled with surpluses in their balance of payments. The peak in oil revenues was reached in 1980. Thereafter, oil revenues fell considerably. The decline in real terms of oil revenues was partly alleviated over the 1981–84 period, owing to the appreciation of the dollar and the consequent decline in oil exporters' import costs (Table 28).

As landlord and taxing authority, oil revenues accrue to a host government. The judicious use of these revenues therefore is incumbent on the

Table 28 Financial data on oil exporters, 1972–84

	Oil revenues(1) ($ billion)*	Current account balances(2) ($ billion)*	Oil exporters' import costs index 1977 = 100
1972	14.5	3.0	
1973	36.0	7.0	
1974	109.0	59.0	
1975	100.0	27.0	
1976	126.0	36.0	
1977	138.0	29.0	100.0
1978	133.0	4.0	—
1979	197.0	62.0	—
1980	279.0	113.0	149.0
1981	253.0	47.4	144.2
1982	201.9	−17.3	138.4
1983	160.4	−20.7	135.6
1984	159.4	−6.8	130.4

* Current dollars

Sources: (1) Shell Briefing Service, *Energy in Profile*, London-Rotterdam, No. 5, 1983, p. 5; and *Oil, Gas and Coal in 1984*, No. 2, 1985, p. 6. (2) Based on IMF's 'Directory of Trade Statistics' for OPEC plus Ivory Coast, Malaysia, Mexico and Oman: Philip K. Verleger, *Petroleum Economic Monthly*, Charles River Associates, Boston, June 1985, cited in *Middle East Economic Survey*, 22 July 1985, pp. A4–A5.

government of the country concerned. Performance depends accordingly on the wisdom of political leadership in the choice of strategies and projects, as well as on the efficiency of the civil service in administering government policies and decisions.

Analysts are interested in ascertaining whether oil-exporting countries have put the opportunity of a surge in oil revenues over 1974–80 to good use. A general answer is bound to be too simplistic, in view of the different circumstances of the countries concerned. The following section attempts to identify major problems encountered by oil-exporting countries in the productive use of their oil revenues.

National Priorities and Strategies

Developmental projects have largely been orientated towards three areas: infrastructure, industrialisation and social welfare. Metropolitan areas have often obtained precedence over rural areas in budgetary allocations. Emphasis has often been put on large or 'show' projects whose beneficial effects on society have proved distant or doubtful. Lack of care has often led to redundant facilities or unfinished projects, with resources going to waste and business confidence shaken.

Inflationary pressures were one of the by-products of mistaken development, which was often due to the crowding-up of construction or industrial projects (some of doubtful economic value) over relatively short periods of time. This concentration provoked bidding-up of scarce resources and the straining of infrastructural facilities. Scarce resources, bottlenecks and inelasticities of domestic supply have tended to encourage consumers to satisfy their demand through imports, in the absence of trade restrictions. Rising domestic costs were also the result of padding prices by intermediaries whose utility to consumers was questionable (see section below, 'Transaction Mores').

Governments have sought to counter the evils of inflation by various means, the effectiveness of which have depended on the country's resources and on the international environment. Among these strategies are: open competitive trade, productive investments, infrastructure investments, reduced administrative and bureaucratic procedures, slowing down of public expenditure subsidies for essential goods and services (in particular food, housing, fuel, education and medication). The impact of such strategies on moderating inflationary pressures has been noteworthy over the 1970s and 1980s in certain Gulf states.

The responsibility of national governments for the mismanagement of their oil wealth has been candidly admitted by some of their officials. The public authorities of several oil-exporting countries failed to protect their people against the evils of inflation and of dislocations in the structure of the economy. As a result, the productive capacity of many sections of the population was sapped. As the secretary-general of OAPEC and former minister of planning in Libya has clearly stated, many oil nations became saddled with dramatic socio-economic problems:

> Let us return to the impact of oil on internal economic development. Perhaps the most serious adverse impact of oil-induced inflation and excessive rural or nomadic–urban migration is the destruction of the traditional productive activities of the people, without replacing them with new forms of productive employment. On the contrary, what usually happens is that people who are partly productive in their subsistence economies join the unemployed personnel of the public and semi-public sectors. Those who join the private sector, where such opportunities exist, generally become either traders, speculators, small shopkeepers or just common labourers in the construction sector. In any case, the whole population becomes more one of urban consumers than producers or entrepreneurs. For all of them, oil money provides the easiest and most attractive source from which to increase their income. Of course, those with economic muscle and political power manage to get much more of it than others. But, in all cases, a sharp dichotomy develops between real productive efforts and monetary rewards.[38]

The social costs of development thus arise from misplaced priorities and the misguided use of resources. Whim, prestige and unnecessary projects have had negative psychological effects on a population which lacks some of the basic amenities of modern life: salubrious housing, medical services,educational and training facilities, and rural development—a significant number of the population being rural. Among those projects which have incited criticism are the building of luxurious palaces and clubs, prestige monuments and vast stadia, and the creation of advanced industries with little relation to the material or human domestic resource base, redundant plants and facilities (e.g. the large international airports only a short distance apart in Abu Dhabi, Dubai and Sharjah within the United Arab Emirates which has a total population of about one million).

One Nigerian academic attributed the failure of development in his country to flagrant waste, notably,

> the crass manner in which what is left of the oil revenue has been managed. Many projects are abandoned half-way on the construction sites; completed projects—houses, factories, communication buildings, schools, and hospitals, etc.—are not brought into use because the necessary machines, equipment, or other facilities either are inappropriate or cannot be operated because the skilled personnel are yet to be trained or recruited.[39]

Economic problems, a Nigerian government minister openly admitted, are 'clearly man-made—made in Nigeria. It is the necessary price we have to pay for our lack of vision, our lack of sustaining moral ethos, our avarice, our vanity, our false conception and perception of what constitutes the good life' . . .[40]

Work Conditions

Oil 'wealth' is largely a natural endowment. In so far as some producers have prolific oil deposits, they obtain a surplus equal to the cost differential of their oil versus 'marginal oil' produced from fields which just manage to break even. The fact that some countries derive much of their prosperity from 'unearned' income has fostered a '*rentier* mentality'. The latter has discouraged entrepreneurship and hard work, and has promoted a spirit of easy gain as derived from monopoly positions or safely-held, well-paid jobs.

The *rentier* or easy-gain attitude is not conducive to hard work and initiative. This is notably true for bureaucracy. In oil-rich Saudi Arabia, a recent survey showed that 75 per cent of government employees arrive late for work, 69 per cent leave before the scheduled time, 51 per cent absent themselves without permission and about 17 per cent draw their salaries without working at all.[41]

A rapid increase in oil income has often induced governments to launch ambitious development plans. The speedy implementation of various projects—in the absence of national skills—has compelled a number of countries to use expatriate manpower on a massive scale. This was the case for the sparsely populated oil-exporting countries, such as the Arab states in the Gulf region and Libya. To illustrate, the work-force of the United Arab Emirates was 467,000 in 1980—90.2 per cent of which were foreign, and 9.8 per cent nationals. The figures for Qatar were, respectively, a 92,000 work-force—84.8 per cent foreign and 15.2 per cent national. For the Arab Gulf states (Bahrain, Kuwait, Oman, Qatar, Saudi Arabia and the UAE) and Libya, some two-thirds of the working population in 1980 (a total of 5.9 million) were expatriate.[42] Another source gave details of the expatriate ratio in the work-force of Kuwait, Libya and Saudi Arabia in 1980 at 78, 34 and 53 per cent, respectively.[43]

Benefits from the oil boom have been primarily geared, as expected, to nationals. Foreign residents and their children, even if established for several decades, have often been denied citizenship or permanent residency and the advantages they carry in civil rights, property or business ownership, educational benefits and social welfare. The discrepancies of opportunity between 'first-class citizens' and 'second-class residents' can be a source of latent tension and a drag on performance. According to one Kuwaiti executive, 'granting citizenship to more foreigners would consolidate loyalty to the state, simplify manpower problems, lessen dependence on more immigration and motivate Kuwaitis to work harder'.[44] Naturally, existing citizens do not wish to dilute their political role by promoting a campaign to naturalise foreign residents.

The emigration of foreign workers often means a drain on valuable talent from their countries of origin. This is compensated for, however, by salary remittances to their home countries. Workers leave their home countries for a variety of reasons. Chief among these are the better work opportunities they may have abroad, higher income, greater personal satisfaction or greater sense of political security. Immigrant workers in host countries have borne the brunt of a number of ills: cultural alienation (including the absence of appropriate schooling and leisure), harsh physical environment (i.e. work and living conditions), absence of union rights, exploitation by recruitment agencies, prohibition or costliness of bringing one's family too.

Expatriate labour has also suffered from lack of job stability. Some host governments have not hesitated to terminate the work permits of immigrant labourers abruptly, whenever faced with economic recessions or following on disagreements with the countries of origin of that labour. Nigeria offers an outstanding example of a country which resorted in 1983 and 1985 to massive expulsions of foreign labour—estimated at over 2 million persons—after welcoming them in the boom period of 1974–81. In the grip of a severe recession, the Nigerian authorities did not hesitate to resort to harsh expulsion, ostensibly on the grounds that the foreign workers in question had not regularised their residence and work permits. Nigeria, in expelling foreign workers, who were mostly from neighbouring African countries, did not feel restrained by its membership of the Economic Community of Western African States or its presumed solidarity with other African nations.[45] Libya resorted to the same approach in the summer of 1985 when, faced with a considerable contraction of economic activities, it expelled scores of thousands of foreign workers. Those who bore the brunt of the abrupt termination of work and residence permits were nationals of countries with whom Libya had political differences, notably Egyptians and Tunisians.[46]

Needless to say, even among nationals, equal opportunities for benefiting from the oil boom have not been assured. Some sections of society have lived in relatively underprivileged conditions, have been treated as economic and/or social inferiors by ruling elites, and have been the first victims of economic reverses.

Transaction Mores

Commissions for sponsors are generally viewed in Saudi Arabia and other Gulf states as compensation for actual or potential services. These services include:

- the function of guarantor offered to foreign companies or their employees, as required by laws and regulations; when large government contracts are at stake, sponsors accredited by the authorities are usually highly placed in the social–business pyramid;
- completion of all legal and administrative formalities for the issue by the public authorities of entry visas and residence permits to expatriates;

— introduction to, and support in relation with, decision-makers of the public or private sector—especially with a view to negotiating and obtaining business contracts, as well as ensuring their execution in conformity with the law.

The Saudi authorities apparently view these commissions (which have averaged some 5 per cent of the value of contracts or of companies' turnover) as:

— a partial compensation for the social/administrative responsibility delegated by the authorities to the sponsor who has to screen, scrutinise and vouch for the activities of individuals and foreign companies in Saudi Arabia;
— one additional means of tapping (in favour of nationals) attractive profits generated by foreign companies' activities in Saudi Arabia.

Commissions need not, however, necessarily constitute a levy on profits. They are likely to permit 'padding' of costs which are passed on in higher prices to the consumer. Examples of excessive commissions and favouritism abound. Among the cases which received publicity in the petroleum field is that of the Italian state-owned ENI, which contracted in 1979 to buy 12.5 million tons of crude oil from Petromin, the Saudi state-owned oil company. The purchase price, inclusive of commissions (estimated at $120 million, or 7 per cent of the purchase price of crude) was nevertheless reported to have come out at below the market price.[47]

In some emerging countries it is not easy to distinguish the domain of the public Treasury from that of the privy purse of ruling elites, in the absence of clear-cut responsibilities and of the public scrutiny found in Western democracies. In certain communities, a public official who takes his 'commission' from a contract to spread it to his party, his village or his tribe may feel righteous and loyal—an action that would be condemned in another society. Moreover, it is quite customary in some countries for government officials to own and oversee business interests in the same area of their public life—a practice considered quite improper in other countries.

Some individuals, whether government or private employees, have resorted with impunity to the circumvention or violation of laws and regulations. A Nigerian petroleum minister frankly stated that 'Quick money-making devices are often resorted to by quite a good number of Nigerians. Illegal oil deals, illegal bunkering, smuggling, unauthorised crude-oil allocations, fronting, con men, etc., all conspire to disrupt and even destroy the economy'.[48]

Reliable researchers have investigated the extent of corruption in a number of countries, and the role of those in public office who use or allow others to use their position for illicit private gain. One study states:

> This is a common practice in the Gulf, but it incurs resentment when it becomes excessive and the profits too large. . . . Occasionally, there have been arbitrary awards or contracts of inflated budget allocations to sectors associated with particular members of ruling hierarchies.[49]

Nationals are often critical of Western companies for fostering the atmosphere in which their officials are corrupted. Saudis:

> talk of large payments being offered to deputy ministers to secure contracts. It is widely assumed that most large business transactions are accompanied by some side payments or commissions. Those Saudis who bemoan the demise of traditional values place more of the blame on the companies who tempt officials to win their favour than on the Saudis who succumb to the lure of enriching themselves at no effort and little risk.[50]

One publicised case of corruption in the oil industry was that relating to Pertamina, Indonesia's state oil company. Western enterprises or intermediaries and national officials have been involved in multimillion-dollar bribes and thefts over the period 1968–1980.[51] Elsewhere wide publicity was given to the corruption within Mexico's state oil monopoly, which took place between 1976 and 1982. Its president, then Diaz Serrano, embezzled $34 million and indulged in lavish spending of public money on the powerful and corrupt oil workers, union. To get jobs, workers had to pay union leaders a fee, varying in size in relation to the importance and tenure of the position sought. The union itself owns farms, oil-contracting companies, aircraft and stores—with the right to assign 50 per cent of all Pemex contracts.[52]

Corruption—admittedly difficult to ascertain—carries with it a social stigma. The penalty could reduce the value of corrupt transactions. Whether by evolution or revolution, the fruits of corruption are eventually taxed, confiscated or recuperated.

Industrial nations cannot be indifferent to factors of waste and corruption that exist in the OPEC area. On the positive side, with the growth of these two factors, 'OPEC's income will be recycled to us [the industrial Western nations] even faster'. On the negative side, 'unrestrained corruption would end in a series of catastrophes after the Iranian model', according to a former US ambassador to Saudi Arabia.[53]

More generally, illicit payments made in order to win or keep advantages and privileges have been a practice not uncommon to business—national or transnational—operating under totalitarian and liberal regimes, in East and West, in developed and developing countries. They also exist under different political and economic conditions: boom, recession or stagnation; peace or conflict. Among the social costs of corruption are the following:

— the threat that it could breed general dissatisfaction, which could produce tension, extremism, strife and the overthrow of rulers and the socio-political order;
— public officials tend to block transactions unless they obtain the required bribes, and many people are unproductively employed in securing the favours of these officials;
— perpetuation of individuals in power, despite their poor or even bad performance;
— stimulation of wasteful and illegal practices such as the evasion of taxes, illegal export of capital and improper awards of projects.[54]

Judged from the standpoint of expediency, the use of 'sweeteners' in business transactions could occasionally be positive; for example, if national productivity and general welfare improved as a result. Otherwise, they are wasteful. It should be noted that analysts in this field encounter problems of objective assessment, including the problem of leaving it to private economic agents to decide on the nature or the scale of such 'sweeteners'.

It is reported that 'in many countries, not only are foreign payments not outlawed, but they are encouraged'. These countries comprise Germany, Canada, France, Japan and the United Kingdom.[55] In 1977 the United States introduced the Foreign Corrupt Practices Act to prohibit American companies from bribing foreign officials. The corporation violating the law is liable to a $1-million fine, and the individual commiting the violation is subject to a $10,000 fine or a five-year prison sentence, or both. The Act allows for authorised bribery (known as grease payments or gratuities) 'to get low-level bureaucrats to perform their duties of processing forms, approving licences or unloading a shipment of perishable goods from a boat in dock', according to US trade representative William Brock (1981). However, Brock argued that the Act's requirements represented 'one of our nation's most serious export disincentives', especially in developing countries which 'are the growth markets of the future'.[56] Human ingenuity can circumvent stringent laws, including the US Foreign Corrupt Practices Act. In Indonesia, so-called charity contributions were used by US companies as a cover to win major contracts.[57]

Harsh law enforcement in OPEC countries has had a deterrent effect. Thus, the South Korean firm Hyundai Construction was banned in April 1980 from operating in Saudi Arabia and was fined $900,000, following an attempt by a company executive to bribe a government official. The executive in question was furthermore sentenced to thirty months in prison.[58]

For the long history of bribery, starting with Pharaonic Egypt, the reader is referred to two interesting works on the subject.[59] Rather than wide-ranging controls, corruption can be more effectively 'limited by avoiding administratively created scarcities; by reducing controls on international trade and payments; and by improving the accountability of officials in the areas where regulations or administrative discretion remain'.[60]

All in all, developmental challenges encountered by oil-exporting countries are many and complex. One can, nevertheless, identify five areas of general concern:

— structural problems, essentially in the form of inadequate or inappropriate infrastructure facilities;
— legal handicaps which preclude the efficient employment of domestic or foreign resources (e.g. absence of adequate guarantees to private capital);
— bureaucracy, which could unduly hamper or delay decisions on economic projects;
— political interventionism or arbitrariness, which could thwart contracts and upset *bona fide* investors;
— general economic mismanagement, often appearing in the form of waste, misplaced priorities and inflation.

Security Burden

Oil-exporting countries have spent significant portions of their GNP on armaments and military personnel. This is particularly true of Middle Eastern countries, whose military arsenals have been accumulated at great expense, with the two superpowers acting as the major purveyors, followed by France and the United Kingdom. Spending in this area by Arab oil-exporting countries reached $361 billion in 1974–83, or about 32.5 per cent of the oil revenues received during that period. In terms of GNP, total Arab countries' military spending was equivalent on average to 14 per cent over 1979–83, although there are wide variations among countries and over years.[61]

According to the Stockholm-based International Peace Research Institute, Middle East countries accounted for 48 per cent of the developing nations' total purchase of arms over the 1970s. Their purchases rose from $11.2 billion in 1971 to $30.3 billion in 1975 and $37.9 billion in 1980. Saudi Arabia's military budget is among the largest in the Middle East and was estimated at about $40 billion for 1981.[62] Its purchases of military equipment from the United States over 1950–82 reached $23.3 billion. This excludes Saudi underwriting of American military expertise, which involved, early in 1983, some 5,300 US military personnel.[63] As of 30 September 1979, Saudi Arabia placed $22.2 billion of undelivered foreign military sales orders with the United States. Money needed to pay for these orders, including potential terminal costs, was deposited in a trust fund with the US Treasury.[64] Other big military spenders in the Gulf were Iran and Iraq.

Western officials and executives have argued that military sales are one means of recycling to the industrial countries the export earnings of oil-exporting developing countries. The president and director-general of the French state-owned oil group Elf-Aquitaine (and formerly a Gaullist minister of defence) expressed his concern regarding the impact of declining oil prices openly: 'In the intermediate term, the [oil-]producing countries which are our big clients for infrastructure and armament goods, as is the case for example with Middle Eastern countries, could reduce their purchases from France in view of their declining resources.'[65] France exported to Iraq in 1981 and 1982 military equipment worth 28 billion frs., and as much again of civilian equipment. Iraq and Saudi Arabia alone absorbed some 70 to 80 per cent of French armament exports in 1982–83 and largely contributed to ensuring the functioning of an industry which employs, directly or indirectly, one million French workers. To finance its ongoing war with neighbouring Iran, begun in September 1980, Iraq had by early 1983 exhausted its accumulated financial reserves and $35 billion granted by Arab Gulf states (i.e. Saudi Arabia, Kuwait and UAE). The French Government had also contributed some 15 billion frs. in export credits to Iraq.[66]

US exports of military goods and services have, like those of France, largely been concentrated in the Middle East since 1974 with the rise in oil-export earnings. Outside Israel, two OPEC countries (Iran and Saudi Arabia) and one OAPEC member (Egypt) were the principal importers:

- Iran's military imports from Western countries rose from $34 million in 1973 to $2.4 billion in 1977, but declined drastically by 1980 after the demise of the Shah and the establishment of the Islamic regime;
- Saudi imports of US military equipment rose sharply to account, in 1982, for $4.4 billion or 36 per cent of total, fully-paid-for US military sales (i.e. excluding the US grant-aid programmes);
- Egypt was the second-largest US customer in 1982, with imports of $1.2 billion.[67]

Expenditure on armaments has been at the expense of civilian investment and has often led to an arms race. Moreover, the temptation to solve internal or external political problems by the use of military force has been great. The military 'solution' has often been resorted to, the most dramatic cases of this being the Iraqi–Iranian and the Arab–Israeli conflicts. Furthermore, 'military might' is no assurance of internal political stability. Regimes have been toppled despite the fact that they appeared to be based on sophisticated, solid military-security structures.

The demise of rulers and changes of regimes can well be attributed in large part to blunders in public policies. Neglect of large sections of the population, rural or urban, has been a source of social tension and upheaval. Some analysts consider that Shah Modammed Pahlavi lost his throne because of the disaffection of many of his people, who had not derived any benefit from his grandiose schemes and who resented his autocratic exercise of power.

In the mid-1970s some major oil-exporting countries 'could afford both guns and butter—rapid increases in military spending and extensive social provisions'.[68] With declining oil revenues and current account deficits in the 1980s (see Table 28), these countries' military spending fell significantly. Nevertheless, a few of these countries (such as Saudi Arabia) still had some of the highest figures for per capita military spending in the world.[69]

The relationship between development and security is manifest in influencing the production of oil resources and the use of oil revenues. Military budgets compete with civilian projects in their claims on oil revenues. Security, however, has not only a military content. Indeed, a nation's wealth judiciously used in the civilian sector should contribute to improving the economic security of the oil-exporting countries concerned. Economic security, an essential component of overall security, has domestic as well as transnational dimensions. At the domestic level, economic security consists in assuring sustained growth in national prosperity, to be shared by large sections of the population. At the transnational level, it consists largely in attaining a wide diversification in trading partners (suppliers and customers), as well as in enhancing the versatility of the national economy and its capacity to respond readily to changes in world supply and demand conditions. Economic security, thus defined, is an important line of defence against potential social unrest and internal destabilisation, and against vulnerability to external economic reverses.

NOTES

1. See, for example, E. J. Mishan, 'GNP—Measurement or Mirage?', *National Westminster Bank Quarterly Review*, London, November 1984, pp. 2–13.
2. For example, GNP statistics cannot readily assess changes in the progress of scheduled castes—known as 'harijans' (formerly 'untouchables')—in India. These have traditionally suffered murder, maiming, dispossesion and humiliation. See Minority Rights Group, cited in 'An upward sweep is subversive', *The Economist*, London, 2 April 1983, p. 80; and 'Inde-Intouchable . . . par mariage', *Le Monde*, Paris, 11 October 1985, p. 7.
3. See Manouchehr Takin, 'Recoverable oil and strategies for production', *OPEC Bulletin*, Vienna, September 1984, p. 14.
4. See 'Nigeria's Oil: Flaring off the Future', *African Business*, London, October 1984, pp. 16–19.
5. The World Bank, *The Energy Transition in Developing Countries*, Washington DC, 1983, pp. 36–7.
6. 'Nigeria: NNPC', *OPEC Bulletin*, Vienna, July/August 1985, p. 29.

7. 'Saudi Arabia Sweetens Terms for Petrochemical Ventures', *Middle East Economic Survey* (*MEES*), 13 December 1983, p. A8.
8. *OAPEC Bulletin*, 'Editorial', Kuwait, April 1985; and Khalid Al-Shawi, 'The Art of Negotiation and the Arab Gulf States' Ability to Negotiate the Export of Their Petrochemical Products', ibid., July 1985, pp. 8–16.
9. 'Petrochem talks stall on EEC Tariffs', *Middle East Economic Digest* (*MEED*), Special Report, *Saudi Arabia*, July 1985, p. 55.
10. The Arab–British Chamber of Commerce, 'Saudi Arabian Petrochemical Exports: The EEC Protectionist Measures and SABIC's Prospects', in *Arab Gulf Journal*, London, April 1985, pp. 9–20.
11. See, for example, United Nations Industrial Development Organisation, *Prospects for Industrial Development and for a Capital Goods Industry in Indonesia*, Vienna, 20 July 1984.
12. Yannis A. Stournaras, 'Is the Industrialisation of the Arab Gulf a Rational Policy?', *Arab Gulf Journal*, London, April 1985, p. 26; see also United Nations Industrial Development Organisation, *Energy-Intensive Industrialization in the Middle East: Third World Opportunities*, by Thomas R. Stauffer, Vienna, 28 August 1984.
13. Tam David-West, Minister of Petroleum and Energy, 'Oil Politics and Development in Nigeria', *OPEC Bulletin*, Vienna, May 1985, p. 14.
14. See, for example, 'Oil and Agriculture: The Nigerian Experience', in *World Development Report 1982*, The World Bank, Washington DC, p. 46.
15. *Business Week*, New York, 27 February 1984, p. 41.
16. Saudi Monetary Agency, *Statistical Summary*, Riyadh, 1984, pp. 33–4.
17. T. W. Lippan, 'Saudis get US pledge to honor food deals', *International Herald Tribune* (*IHT*) and *Washington Post*, 11 May 1983.
18. Saudi Arabian Monetary Agency, *Annual Report 1404 (1984)*, Riyadh, pp. 84–91; and 'Green grow the deserts O', *The Economist*, London, 6 April 1985, pp. 74 and 75.
19. See Gerald F. Seib, 'Libya's ambitious water pipeline project draws heavily on expertise of US firms', *Wall Street Journal*, New York, 11 October 1985, p. 2; and 'Great Man-Made River Shortlist Due in November', *MEED*, 21 September 1985, pp. 23–4.
20. See Alan Berg, *Malnourished People—A Policy View*, The World Bank, Washington DC, June 1981.
21. See, for example, 'The Worm that Weakens', *Developmental Forum*, Geneva, July–August 1981, p. 13.
22. Soedjatmoko, Rector of the UN University, 'Inseparable partners: The inter-linked futures of food and energy', *Developmental Forum*, Geneva, June 1984, p. 6.
23. US General Accounting Office, *Importance and Impact of Federal Alcohol Fuel Tax Incentives*, Washington DC, 6 June 1984, pp. 1–2 and 37.
24. Ibid., p. 9.
25. Alan Riding, 'Brazil stresses its oil success', *New York Times*, 16 April 1984, pp. D1 and D8.
26. Patricia Boyce and Hayden Llewellyn, *World Trade Distortions*, Melbourne, AIDA Research Centre, 1983, p. 144.
27. Ernane Galvêas, Minister of Finance, *The Brazilian Economy: Challenges and Accomplishments*, Brazilia, Central Bank of Brazil, 1982, p. 13; and Michael Crabbe, 'Brazil—Sustained rise in oil production', *Petroleum Economist*, London, December 1984, p. 454.
28. Roger Cans, 'Brézil—canne à sucre contre piranhas', *Le Monde*, Paris, 2–3 December 1984, p. 6.
29. 'Brazil's bumpy ethanol road', *The Economist*, London, 19 June 1982, p. 88; and 'Brazil's ethanol fuel: a victim of success?', *Business Week*, New York, 24 August 1981, p. 47.
30. See, for example, Michael Barzelay and Scott R. Pearson, 'The Efficiency of Producing Alcohol for Energy in Brazil', *Economic Development and Cultural Change*, Chicago, Illinois, October 1982, p. 144; and Reuter, 'Brazil's fuel programme threatened by oil prices', 2 July 1985, cited by *OPEC Bulletin*, Vienna, July/August 1985, pp. 74–5.

31. Ignacy Sachs, Director of UNU in charge of the Food Energy Nexus Programme, in *Development Forum*, New York, United Nations, July–August 1985, p. 6.
32. See L. R. Brown, 'Competition for Cropland', *Economic Impact*, Washington DC, 1980/4, pp. 32–40; and F. H. Sanderson, 'Benefits and costs of the US gasohol program', *Resources*, Washington DC, Resources for the Future Inc., July 1981, pp. 2–13.
33. See R. Picciotto, 'Malnutrition et développement agricole', *Nouvelles et opinions sur le développement*, Paris, The World Bank, April 1981, p. 1; and US General Accounting Office, *World Hunger and Malnutrition Continue: Slow Progress in Carrying out World Food Conference Objectives*, Comptroller General Report to the Congress, Washington DC, 11 January 1980, p. 22.
34. Food and Agriculture Organisation (FAO) of the United Nations, *Commodity Review and Outlook: 1980–81*, Rome 1981, pp. 42–51 and 59–63.
35. The United Nations University Newsletter, *UNU*, **6**, 1, Tokyo, p. 3.
36. J. D. Miles, 'Alcohol from Grain: A Transition Fuel', *Voice*, Federal Reserve Bank of Dallas, May 1981, pp. 9–10.
37. See The World Bank, *World Development Report*, 1980, Washington DC, August 1980, p. 18.
38. Ali A. Attiga, 'Economic Development of Oil-Producing Nations', *OPEC Bulletin*, Vienna, November 1981, p. 10.
39. Akin Fadahunsi, 'Nigerian Oil Wealth Bypasses Rural Majority', the United Nations University, *Newsletter*, Tokyo, November 1983, p. 7.
40. David-West, op. cit., pp. 16–17.
41. Public Administration Institute (Riyadh), cited by *MEED* Special Report, *Saudi Arabia*, July 1985, p. 63; see also Hazem El-Beblawi, *The Oil Decade: An Appraisal in Perspective*, The Industrial Bank of Kuwait, September 1983, pp. 37–8.
42. *Joint Arab Economic Report, 1983*, published by the Arab Monetary Fund, General Secretariat of the Arab League, Arab Fund for Economic and Social Development, and Organisation of Arab Petroleum Exporting Countries, edited by the Arab Monetary Fund, Abu Dhabi, pp. 70 and 209.
43. Naiem A. Sherbiny, 'Expatriate labor in Arab oil-producing countries', *Finance and Development*, Washington DC, IMF/World Bank, December 1984, p. 35.
44. Suleiman Mutawa in 'Kuwaitis are concerned with pressures to broaden rights of foreign workers', *IHT*, 16 March 1981, p. 6; see also Saad Eddin Ibrahim, *The New Arab Social Order: A Study of the Social Impact of Oil Wealth*, Boulder, Colorado, Westview Press, and London, Croom Helm, 1982, 208 pp.; and Gilbert Beaugé, 'Les migrations internationales au Moyen-Orient', *Revue Tiers Monde*, Paris, July–September 1985, pp. 485–679.
45. See, for example, 'Nigeria—go away again', *The Economist*, London, 4 May 1985, p. 42.
46. See *MEED*, 7 September 1985, p. 20; and Michael Deuré, 'Tunisie-Libye: les frères enemis', *Le Monde*, 8–9 September 1985, pp. 4–5.
47. According to Giorgio Mazzanti, the then president of ENI: see, for example, 'Italian Company's Oil Supplies Halted', *MEED*, London, 7 December 1979, p. 53; 'ENI "Affair" Creates Political Unrest in Italy and Saudi Arabia', *Arab Oil and Gas*, Paris, 16 December 1979, pp. 8–9; and 'Italian Minister Visits', *MEED*, London, 22 February 1980, p. 32.
48. David-West, op. cit., p. 14.
49. Valerie Yorke, *The Gulf in the 1980s*, The Royal Institute of International Affairs, London, 1980, p. 19.
50. See William B. Quandt, *Saudi Arabia in the 1980s—Foreign Policy, Security, and Oil*, Washington DC, The Brookings Institution, 1981, p. 148.
51. 'L'affaire Pertamina—quand le général Sutowo se fait restaurateur', *Le Monde*, Paris, 5 February 1977, p. 34; and 'Pertamina Thefts Revive Corruption Fears', *IHT*, 9 February 1981, p. 9.
52. See, for example, Alan Riding, 'Mexico Takes Up Challenge of Pemex', *IHT*, 26–

27 February 1983, pp. 7 and 9; and 'Mexico—the big fish', *The Economist*, London, 9 July 1983, pp. 52 and 55.

53. James E. Akins, 'Can International Energy Supplies Support the United States' Energy Needs?', in *The International Essays for Business Decision-Makers*, vol. IV (ed.) M. B. Winchester, Houston, Texas, Center for International Business, 1979, p. 67.
54. See 'Corruption' in World Bank, *World Development Report 1983*, New York, Oxford University Press, 1983, p. 117.
55. J. Kaikati and Wayne A. Label, 'The Foreign Antibribery Law: Friend or Foe?', *Columbia Journal of World Business*, New York, Spring 1980, pp. 49.
56. 'Brock Criticizes Bribery Law', *New York Times*, 17 December 1981, D13.
57. 'A pint of blood? or $200?', *Forbes*, New York, 12 October 1981, p. 96.
58. *MEES*, Nicosia, 12 May 1980, p. 17.
59. For example, Majid Tobia, 'Bribery has its History', *Al'Arabi*, Kuwait, Ministry of Information, February 1981, pp. 102–6 (monthly bulletin, in Arabic); and John T. Noonan, jun., *Bribes*, New York, Macmillan, 1984.
60. 'Corruption', in *World Development Report, 1983*, op. cit.
61. See 'Arab Oil-Producing Countries have spent $361 Billion on Defense in 10 Years', *Arab Oil and Gas*, Paris, 16 September 1984, pp. 26–9.
62. *Arab Oil and Gas*, Paris, 16 November 1981, pp. 12–13.
63. R. Halloran, 'Saudis consider buying new US arms', *New York Times*, 9 March 1983, p. A3.
64. Comptroller General of the United States, *How Military Sales Trust Funds Operate: Saudi Arabian and Iranian Funds Compared*, Washington DC, 28 January 1981, pp. 1–2.
65. Albin Chalandon, 'L'OPEP doit négocier avec les autres pays producteurs', *Le Monde*, Paris, 27 January 1983, p. 27.
66. 'Un soutien risqué', *Le Monde*, Paris, 5 February 1983, p. 1.
67. 'US Exports of Military Goods and Services', *International Letter*, Federal Bank of Chicago, 21 October 1983, pp. 1 and 2.
68. Stockholm International Peace Research Institute, *World Armaments and Disarmament—SIPRI Yearbook 1985*, London and Philadelphia, Taylor and Francis, 1985, p. 227.
69. Ibid., pp. 264–5.

7 Conclusion

Oil is an eminently transnational industry, with its raw materials available for export in substantial quantities in a few countries, while consumption needs are concentrated in a few other countries, and with a large number of business groups with oil activities in more than one country. Moreover, during a large part of the twentieth century, oil has been an important input in economic, social and technological performance, as well as a powerful factor in transnational affairs, considerably affecting trade, finance and inter-state relations.

Rapid economic growth in energy-importing countries has been boosted by abundant oil supplies, competitively low-priced (notably over the 1950s and 1960s), while surging oil prices (e.g. in the 1970s and early 1980s) dampened world economic activity. Rising revenues of oil-exporting countries, put to judicious use, should enhance the development of their domestic economies. This opportunity has not, however, always been optimally exploited, as we have seen.

Notwithstanding basic changes during the history of the world oil industry, the foregoing analysis has shown that two major groups of protagonists have significantly influenced oil events: major companies and some governments. Their respective roles are reflected on further in this conclusion.

OPERATORS' BEHAVIOUR AND COMPETITIVE FORCES

Historical evidence has shown that a concentrated market structure in the petroleum sector has generally facilitated moves towards *entente* (overt or covert) and to co-operation among participants with a view to exploiting market positions in order to attain certain perceived advantages. Thus, their limited numbers have tempted sellers into forming producers' associations or cartels in order to protect or increase their profits. Symmetrically, petroleum importer countries' search for concentration proved possible, thanks to the thrust of a few major countries.

However, concentrated markets, restraint on trade and attractive performance have, in due course, enticed outsiders to join in. Thus, prompted by attractive returns or supported by their governments, newcomers expanded their share in the production, transport, refining and marketing of petroleum—at the expense of established enterprises. To underpin their market power, the majors have adopted various approaches. The strategy of vertical integration (from wellhead to petrol station) has proved attractive to oil companies, unless arm's length transactions with independent parties can be readily made at lower cost than similar transactions with affiliated companies. Indeed, over the longer term, a measure of vertical integration and geographical spread appear on balance to yield several advantages to operators: security of flows, rationalisation of operations and avoidance of undue dependence on volatile supplies or crude-oil spot prices. To illustrate the drive for vertical integration and geographical spread, one could refer to the cases of the Standard Oil groups of John D. Rockefeller in the late nineteenth century, Royal Dutch/Shell at the start of the twentieth century and Chevron/Gulf in 1984. Major oil companies have also sought to diversify outside the petroleum and energy sector, especially when their liquidity and profits have boomed. However, this diversification has had a mixed impact, and a number of companies have been compelled to divest activities which compromised their profitability.

The oil market does not follow the textbook economic efficiency model. Several imperfections exist, notably in the form of restrictions on free entry, inaccessibility of key information, and inequality in bargaining power. The end result of these imperfections is a measure of market advantage for sizeable, well-established groups *vis-à-vis* smaller or newcomer groups. By exploiting to their advantage dominant market positions, companies and governments cannot, however, expect to derive indefinitely the benefits they seek. This study has illustrated this situation by reference to OPEC countries. In the 1970s these countries, individually and collectively, overshot the price level which would have maintained their dominant position over the medium and longer terms in the world oil market. Thus, OPEC's production stood at 53.5 per cent of the world total in 1972; it fell to about 30 per cent of the world total in the mid-1980s, with oil losing its share of the global energy market in favour of substitutes.

These factors led to excess capacity (downstream and upstream) in the world oil industry and heightened the intensity of competition in the 1980s. This, in turn, has reversed the upward price trend of the 1970s, starting with 1981; the downward price trend of 1981–85 reduced operators' incentives to explore for, and develop, additional energy supplies. Such a downward price trend, if it persists, could put out of business some of the higher-cost energy projects and hence raise consumers' dependence over the longer term on the large and lower-cost oil reserves—mostly located in the Middle East.

On the basis of past experience one can argue that price-fixing over relatively long stretches of time in the world industry should prove increasingly difficult in a market characterised by:

— a large number of operators (brokers, traders, speculators, state companies, independents, as well as majors or mini-majors) all competing at various stages of the oil industry;
— uncertain demand (of final consumers or of stocktakers) and supply conditions, or fluctuating expectations;
— greater use of shorter-term spot contracts, in lieu of stable long-term business arrangements.

The decade of the 1970s is likely to prove, in a historical perspective, a watershed. In that period, oil-exporting developing countries had wrested the initiative from their concessionaires, the transnational majors, in influencing their price and export levels. With a favourable economic and political environment, governments of oil-exporting countries then managed to exploit market conditions, and in particular the inelasticity of demand over the short term. Crude-oil prices rose dramatically twice in the 1970s following on the heels of crises and interruptions in supplies from the Middle East, which hold (and continue to hold) the world's largest concentration of commercial proven reserves: the 1973–74 price increases followed on the Arab–Israeli war of October 1973 and its aftermath, and the 1979–80 price increases accompanied disruptions caused by the Iranian revolution. The second major price surge, topping the earlier one, reached such proportions that it contributed to the slowing of world activity and to a levelling off of world demand for energy in the 1980s. The demand for OPEC oil contracted considerably in that period since:

— oil companies and consumer countries sought to develop supplies outside the OPEC area;
— oil consumers and manufacturers of oil-fuelled equipment found it commercially attractive to invest in conservation and in improved energy efficiency of equipment;
— certain industries managed to substitute other sources of energy for oil.

GOVERNMENT POLICIES

Government policies (and policy changes) in the energy field are often determined by dominant interest groups and their perception(s) of national welfare. Generally, countries are primarily concerned about protecting and increasing their socio-economic prosperity and political stability. For these fundamental concerns, interest groups try to influence government policies. Their influence would affect *inter alia*, sources, volumes and mixes of energy production and consumption, activities of enterprises (private or state-owned), fiscality and regulations, competition, and inter-government relations.

Market economies cannot attain optimal efficiency (also known among economists as Pareto efficiency) unless certain conditions prevail. Chief among these conditions is the existence of perfectly competitive markets. This implies that:

— enterprises acting separately are too small to influence market prices;
— free entry and exit of enterprises is unhindered;
— customers and operators have ready access to full, truthful information on goods and services produced.

The international petroleum industry certainly does not meet this model of perfect competition. Governments have intervened ostensibly to remove impediments to efficient market functioning. Governments have also increasingly responded to public demands for cutting down on external and social costs. Among the latter are health, safety and environmental hazards. Accordingly, government prescriptions aim at obligating the enterprise or the user to assume the costs of eliminating or minimising health and environmental damage to society. Governments have also sought to maximise external benefits from the enterprises' operation; for example, by allowing local communities access to the infrastructure (roads, ports, etc.) originally built by the petroleum industry for its own operations.

Host governments have also been concerned with optimising their profits (in money or kind) from oil operations, as well as with tuning the depletion of petroleum deposits (and other exhaustible resources) to social (as distinct from private) needs. Moreover, governments have generally intervened to 'insure' or 'protect' national security (e.g. by strategic stockpiling) whose value cannot be solely ascertained by reference to oil acquisition costs.

Price control has been resorted to by various public authorities, with mixed results. Some analysts and policy-makers have called for the pricing of natural crude-oil near the replacement cost of close substitutes (e.g. synthetic oil derived from tar sands, coal, etc.) available in large quantities. In the calculation of this replacement cost, the crucial question to be answered is: 'What timespan should be considered?' Short- and longer-run effects are quite different. Thus a sharp increase in oil prices, to match them with higher-cost substitutes, will result, in the short run, in an unacceptable boost to inflation and unemployment, with little additional energy supply forthcoming. In the longer run, demand and supply elasticities are significantly greater. This means that higher energy prices today could well lead in the longer term to excessive supply in relation to current demand, and consequently to an eventual sharp decline in prices. This, in turn, could discourage the continued development of energy supplies.

One policy implication of the previous scenario is that volatile and short-term market forces have to be controlled by public authorities to avoid excessive price fluctuations. To ensure an efficient adjustment of supply and demand over the longer term, the authorities should therefore allow prices to respond to fundamental changes in cost conditions and consumers' preferences.

Not content with regulation, governments in market economies have created state enterprises as an effective tool of public policy. A recommendation for the direct intervention of public authorities is based on the hypothesis that they are better informed about present and future conditions than the sum of private economic agents in an open competitive market. The assumption that governments always have superior knowledge, as compared to market signals, is questionable—as shown by available evidence.

Oil has also been used, along with other goods and services, as an instrument of foreign policy, with various parties threatening or actually denying supplies or blocking outlets. The effectiveness of this instrument is limited in time. The targeted countries can in fact be provoked into hardening their positions, and/or espousing alternative longer-term arrangements to reduce their vulnerability. The costs of such sanctions fall on both contending parties, without an assurance of satisfaction of foreign policy goals for either. Sanctions are generally more effective as threats, to be used with care, rather than to be applied on a continuous basis.

Government policies' effectiveness could be frustrated by lack of perseverence and coherence. Indeed, different departments and government institutions of a given country may, at their own discretion, do different things in the light of what makes economic or political sense to them at the time. Changes in economic and political circumstances also produce new policy orientations.

COMPLEX INTERDEPENDENCES

Oil has been a key factor of interdependence—notably in trade, finance and political relations. Thus, a large price change (increase or decrease) is bound to considerably affect countries and financial intermediaries. The oil price increases of the 1970s were one of the major sources of the expansion of oil-importing countries' foreign debt, while the oil price reductions of the 1980s aggravated the financial position of oil-exporting countries. On balance, analysts recognise that a reduction in oil prices from their peak level of 1981 should be in the interests of world economic growth. However, what should be the optimal price level remains a begged question.

Past events in the world oil industry have highlighted the interdependence of the protagonists concerned. In a multi-polar world, whose prosperity and security is considerably influenced by such a transnational commodity as oil, each group has to tread cautiously if it is not to provoke a drastic reaction from other actors, to the detriment of all parties concerned. This applies to various areas of decision-making, such as prices, trade and investment.

Also of significance in this interaction is the fact that each group is not composed of homogeneous members. There are wide disparities in conditions, needs and priorities. This diversity has furthermore facilitated the development of cross-group relations and alliances. The development of such relations can be seen between the United States and Saudi Arabia. In the 1970s and 1980s both represent major actors in the world oil industry—at the levels of internal market size, know-how and economic–political power for the United States, and at the level of vast commercial reserves for Saudi Arabia. The two countries, furthermore, broadly share the same view of East–West relations.

Generally speaking, industrial oil-importing countries (e.g. the United States) enjoy higher standards of living and know-how, as compared with oil-exporting developing countries. Furthermore, their diversified economies have enabled them to overcome adversities more readily than developing economies. The latter—even those with great oil wealth (e.g. Saudi Arabia)—are comparatively more vulnerable. With small, less diversified economies and lower levels of sophistication, the oil wealth of developing countries is not readily convertible into other sources of productive asset. Indeed, wealth and sustainable progress lies less in the endowment of natural resources than in the systematic continuous development of human talents and expertise in innovation and hard work and in the successful creation of appropriate and efficient institutions.

PROSPECTS

Despite the general quest for stability, producers and consumers have to reckon with the continual dynamism of the world energy market. Protagonists therefore have to adjust to variations or fluctuations in crude-oil and product prices, as well as in production levels. Flexibility in prices and outputs is warranted to permit continuous adjustments to the ever-evolving conditions of:

— global supply and demand for energy;
— the structure of the market and the behaviour of operators;
— consuming or producing countries' policies.

Underlying OPEC countries' search for market stability is their concern for stability in their oil export profits. These countries' heavy dependence on oil revenue renders them anxious to protect their earning capacity, in order to defray their import bills for necessities, especially in periods of economic difficulty. Such stability in oil revenues cannot, however, be established without the effective co-operation of major oil importers, mainly the leading industrial countries grouped within the IEA.

In the buyers' market of the 1980s and the prevailing price weakness, IEA's energy-importing industrial countries refused an institutional multilateral dialogue with OPEC aimed at promoting the stabilisation of makets. This attitude of non-institutional dialogue with the 'other side' is similar to that espoused by OPEC countries in the sellers' market of the 1970s.

Looking into the longer term (e.g. until the late 1990s), OPEC oil production could well show an upward trend again—assuming no increase in the level of real oil prices from their 1985 level. This is owing primarily to the fact that non-OPEC commercial oil reserves (which are smaller and more costly to develop and produce) are being depleted at a higher rate.

Non-oil producers of energy will have to resolve certain problems if they are to improve their competitiveness *vis-à-vis* oil and gas: for coal, it is production methods and the acid-rain question; for nuclear power, it is nuclear waste and operational safety. For other potential sources of energy, like solar power, the horizon of large-scale commercial production is distant and calls for substantial investments in research and development.

Most parties in the world economy—investors and producers, as well as users—are anxious to avert crises and are interested in a measure of stability of flows and prices. While such stability is elusive, countries and companies have some prospect of attenuating fluctuations and shortening their duration. Among the mechanisms which could achieve this objective one can cite the potential role of corporate or government stockpiles, producers' excess capacity, diversification of energy sources and outlets, versatility of energy users in switching among energy products, and producer–consumer contacts.

The history of world oil since 1860 has been punctuated by confrontations and crises, and the management of these crises has often produced palliatives. What is needed are judicious arrangements which could help to avoid these crises or reduce their impact. One such arrangement is to allow for readily available supply capacity (whether through additional oil production from oil-fields or through the release of stored oil) to accommodate unexpected developments, such as interruptions in supplies and sudden increases in demand. Could this be done by a greater measure of understanding, and possibly of co-operation, between producers and consumers? Here lies a major global problem, the solution of which depends on multiple and complex factors affecting international relationships. Speculation about the future is outside the scope of this study. As of the mid-1980s, there is quite a nexus of interests favouring price stability: this includes OPEC countries which are enjoying handsome rent; higher-cost hydrocarbon producers (e.g. North Sea and US producers) and higher-cost, non-oil energy producers (coal, nuclear, etc.); central bankers worried about the possible defaulting of oil-exporting countries (e.g. Mexico) should oil prices collapse; commercial bankers concerned about protecting loans made to oil, gas and other energy projects; and the various industries that service the energy sector. Countries and companies concerned about the longer term have to demonstrate a capacity to adjust readily to changing market conditions, and must be able to develop the skills necessary to tackle the problems that arise in complex and ever-changing socio-political environments. Oil, and energy in general, thus remains a key vector of rivalry or co-operation in transnational relations. Institutions created in response to various challenges posed by the oil sector will have to evolve by adapting their functions and operations to changing realities—if they are to survive and develop. The essence of longevity is adaptability.

Index